THERMOCHEMICAL WASTE TREATMENT

Combustion, Gasification, and Other Methodologies

THERMOCHEMICAL WASTE TREATMENT

Combustion, Gasification, and Other Methodologies

Edited by

Elena Cristina Rada, PhD

AAP | APPLE ACADEMIC PRESS

Apple Academic Press Inc.
3333 Mistwell Crescent
Oakville, ON L6L 0A2
Canada

Apple Academic Press Inc.
9 Spinnaker Way
Waretown, NJ 08758
USA

Exclusive worldwide distribution by CRC Press, a member of Taylor & Francis Group

Printed in the United States of America on acid-free paper

International Standard Book Number-13: 978-1-77188-308-5 (Hardcover)

International Standard Book Number-13: 978-1-77188-309-2 (eBook)

Library and Archives Canada Cataloguing in Publication

Thermochemical waste treatment : combustion, gasification, and other methodologies / edited by Elena Cristina Rada, PhD.

Includes bibliographical references and index.
Issued also in electronic format.
ISBN 978-1-77188-308-5 (hardcover).--ISBN 978-1-77188-309-2 (pdf)
1. Refuse and refuse disposal. 2. Thermochemistry. 3. Incineration.
4. Biomass gasification. 5. Pyrolysis. 6. Hydrothermal carbonization.
I. Rada, Elena Cristina, editor

TD796.7.T54 2016 628.4'45 C2016-900323-X C2016-900324-8

Library of Congress Cataloging-in-Publication Data

Names: Rada, Elena Cristina, editor.
Title: Thermochemical waste treatment : combustion, gasification, and other methodologies / Elena Cristina Rada, PhD.
Description: Toronto; Waretown, New Jersey : Apple Academic Press, 2016. |
Includes bibliographical references and index.
Identifiers: LCCN 2016001420 (print) | LCCN 2016002860 (ebook) | ISBN 9781771883085 (hardcover : alk. paper) | ISBN 9781771883092 ()
Subjects: LCSH: Incineration. | Refuse and refuse disposal. | Sewage--Purification. | Thermochemistry.
Classification: LCC TD796 .R33 2016 (print) | LCC TD796 (ebook) | DDC 628/.74--dc23
LC record available at http://lccn.loc.gov/2016001420

Apple Academic Press also publishes its books in a variety of electronic formats. Some content that appears in print may not be available in electronic format. For information about Apple Academic Press products, visit our website at **www.appleacademicpress.com** and the CRC Press website at **www.crcpress.com**

About the Editor

ELENA CRISTINA RADA, PhD

Elena Cristina Rada, PhD, earned her master's degree in Environmental Engineering from the Politehnica University of Bucharest, Romania; she received a PhD in Environmental Engineering and a second PhD in Power Engineering from the University of Trento, Italy, and the Politehnica University of Bucharest. Her post-doc work was in Sanitary Engineering from the University of Trento, Italy. She has been a professor in the Municipal Solid Waste master's program at Politehnica University of Bucharest, and has served on the organizing committees of "Energy Valorization of Sewage Sludge," an international conference held in Rovereto, Italy, and Venice 2010, an International Waste Working Group international conference. She also teaches seminars in the bachelor, master, and doctorate modules in the University of Trento and Padua and Politehnica University of Bucharest and has managed university funds at national and international level. Dr. Rada is a reviewer of international journals, a speaker at many international conferences, and the author or co-author of about a hundred research papers. Her research interests are bio-mechanical municipal solid waste treatments, biological techniques for biomass characterization, environmental and energy balances regarding municipal solid waste, indoor and outdoor pollution (prevention and remediation) and health, and innovative remediation techniques for contaminated sites and streams.

Contents

Acknowledgment and How to Cite

The editor and publisher thank each of the authors who contributed to this book. The chapters in this book were previously published elsewhere. To cite the work contained in this book and to view the individual permissions, please refer to the citation at the beginning of each chapter. Each chapter was carefully selected by the editor; the result is a book that looks at thermochemical waste treatment from a variety of perspectives. The chapters included are broken into four sections, which describe the following topics:

1. Combustion
2. Gasification
3. Pyrolysis
4. Hydrothermal Carbonization

The articles in each of these categories represent recent important research in thermochemical waste.

List of Contributors

Débora Almeida
Instituto de Macromoléculas Eloisa Mano, Universidade Federal do Rio de Janeiro – UFRJ, Rio de Janeiro, RJ, Brasil

S. Caillat
Université Lille Nord de France, F-59500 Lille, France and Département Énergétique Industrielle, École des Mines de Douai, F-59508 Douai Cedex, France

M. J. Castaldi
Earth & Environmental Engineering Department (HKSM), Columbia University, 500 West 120th street, New York, N.Y. 10027

Howard A. Chase
Department of Chemical Engineering and Biotechnology, University of Cambridge, New Museums Site, Pembroke Street, Cambridge CB2 3RA, UK

D. Chen
Thermal & Environmental Engineering Institute, Tongji University, Shanghai, 200092, China

Ranjana Chowdhury
Chemical Engineering Department, Jadavpur University, Kolkata 700 032, India

Mariusz Dudziak
Institute of Water and Wastewater Engineering, Silesian University of Technology, Gliwice 44-100, Poland

D. Gambier
Maguin SAS, F-02800 Charmes, France

L. Gasnot
Université Lille Nord de France, F-59500 Lille, France and Physicochimie des Processus de Combustion et de l'Atmosphère (PC2A) – UMR 8522 CNRS, Université Lille 1 Sciences et Technologies, F-59655 Villeneuve d'Ascq Cedex, France

F. Le Gleau
Université Lille Nord de France, F-59500 Lille, France, Département Chimie et Environnement, École des Mines de Douai, F-59508 Douai Cedex, France, and Département Énergétique Industrielle, École des Mines de Douai, F-59508 Douai Cedex, France

Y. Ishida
Nippon Steel & Sumikin Engineering Co., Ltd., 46-59, Nakabaru, Tobata-ku, Kitakyushu, Fukuoka, 804-8505, Japan

Manohar Kotha
Department of Electrical and Electronic Engineering, Wollo University, South Wollo, Ethiopia

Su Shiung Lam
Department of Engineering Science, Faculty of Science and Technology, Universiti Malaysia Terengganu, 21030 Kuala Terengganu, Terengganu, Malaysia and Department of Chemical Engineering and Biotechnology, University of Cambridge, New Museums
Site, Pembroke Street, Cambridge CB2 3RA, UK

Michael Langanki
Engler Bunte Institute (Gas Erdol Kohle), Karlsruhe University, 76133 Karlsruhe, Germany

Zhen Liu
National Engineering Laboratory for Coal-fired Pollutants Emission Reduction, Energy and Power Engineering School, Shandong University, 17923 Jingshi Road, Jinan, Shandong 250061, China

Daegi Kim
Department of Environmental Science and Technology, Tokyo Institute of Technology, G5-8, 4259 Nagatsuta-cho, Midori-ku, Yokohama 226-8503, Japan

Chunyuan Ma
National Engineering Laboratory for Coal-fired Pollutants Emission Reduction, Energy and Power Engineering School, Shandong University, 17923 Jingshi Road, Jinan, Shandong 250061, China

Qingluan Ma
National Engineering Laboratory for Coal-fired Pollutants Emission Reduction, Energy and Power Engineering School, Shandong University, 17923 Jingshi Road, Jinan, Shandong 250061, China

X. Ma
Thermal & Environmental Engineering Institute, Tongji University, Shanghai, 200092, China

Maria de Fátima Marques
Instituto de Macromoléculas Eloisa Mano, Universidade Federal do Rio de Janeiro – UFRJ, Rio de Janeiro, RJ, Brasil

Ashmi Mewada
N. Shankaran Nair Research Center for Nanotechnology and Bionanotechnology, SICES College of Arts, Science and Commerece, Ambernath, Maharashtra 421505, India

Neeraj Mishra
N. Shankaran Nair Research Center for Nanotechnology and Bionanotechnology, SICES College of Arts, Science and Commerece, Ambernath, Maharashtra 421505, India

M. Osada
Nippon Steel & Sumikin Engineering Co., Ltd., (Head Office) Osaki Center Building 1-5-1, Osaki, Shinagawa-ku, Tokyo 141-8604, Japan

Xinchao Pan
Institute of Environmental Science &Technology, Hangzhou Dianzi University, Hangzhou 310018, China

Sunil Pandey
N. Shankaran Nair Research Center for Nanotechnology and Bionanotechnology, SICES College of Arts, Science and Commerece, Ambernath, Maharashtra 421505, India

Bhushan Patil
N. Shankaran Nair Research Center for Nanotechnology and Bionanotechnology, SICES College of Arts, Science and Commerece, Ambernath, Maharashtra 421505, India

J-F. Pauwels
Université Lille Nord de France, F-59500 Lille, France and Physicochimie des Processus de Combustion et de l'Atmosphère (PC2A) – UMR 8522 CNRS, Université Lille 1 Sciences et Technologies, F-59655 Villeneuve d'Ascq Cedex, France

E. Perdrix
Université Lille Nord de France, F-59500 Lille, France and Département Chimie et Environnement, École des Mines de Douai, F-59508 Douai Cedex, France

Pandji Prawisudha
Department of Environmental Science and Technology, Tokyo Institute of Technology, G5-8, 4259 Nagatsuta-cho, Midori-ku, Yokohama 226-8503, Japan

Hemlata Sahu
Department of Electrical and Electronic Engineering, Wollo University, South Wollo, Ethiopia

Omprakash Sahu
Department of Chemical Engineering, Wollo University, South Wollo, Ethiopia

Aparna Sarkar
Chemical Engineering Department, Jadavpur University, Kolkata 700 032, India

Sudip De Sarkar
Chemical Engineering Department, Jadavpur University, Kolkata 700 032, India

Madhuri Sharon
N. Shankaran Nair Research Center for Nanotechnology and Bionanotechnology, SICES College of Arts, Science and Commerece, Ambernath, Maharashtra 421505, India

Maheshwar Sharon
N. Shankaran Nair Research Center for Nanotechnology and Bionanotechnology, SICES College of Arts, Science and Commerece, Ambernath, Maharashtra 421505, India

Jing Sun
National Engineering Laboratory for Coal-fired Pollutants Emission Reduction, Energy and Power Engineering School, Shandong University, 17923 Jingshi Road, Jinan, Shandong 250061, China

N. Tanigaki
Nippon Steel & Sumikin Engineering Co., Ltd. (European Office), Am Seestern 8, 40547, Dusseldorf, Germany

Mukeshchand Thukur
N. Shankaran Nair Research Center for Nanotechnology and Bionanotechnology, SICES College of Arts, Science and Commerece, Ambernath, Maharashtra 421505, India

Misgina Tilahun
Department of Chemical Engineering, Wollo University, South Wollo, Ethiopia

Wenlong Wang
National Engineering Laboratory for Coal-fired Pollutants Emission Reduction, Energy and Power Engineering School, Shandong University, 17923 Jingshi Road, Jinan, Shandong 250061, China

Z. Wang
Thermal & Environmental Engineering Institute, Tongji University, Shanghai, 200092, China

Sebastian Werle
Institute of Thermal Technology, Silesian University of Technology, Gliwice 44-100, Poland

Zhengmiao Xie
Institute of Environmental Science &Technology, Hangzhou Dianzi University, Hangzhou 310018, China

Kunio Yoshikawa
Department of Environmental Science and Technology, Tokyo Institute of Technology, G5-8, 4259 Nagatsuta-cho, Midori-ku, Yokohama 226-8503, Japan

Chao Zhao
National Engineering Laboratory for Coal-fired Pollutants Emission Reduction, Energy and Power Engineering School, Shandong University, 17923 Jingshi Road, Jinan, Shandong 250061, China

L. Zhao
Thermal & Environmental Engineering Institute, Tongji University, Shanghai, 200092, China

G. Zhou
National Engineering Research Center for Urban Pollution Control, Tongji University, Shanghai, 200092, China

Introduction

Our world faces enormous challenges. With a growing world population, diminishing fossil fuel resources combined with increasing waste generation will become a critical problem. We need to begin to think in a different way about the materials we no longer use or need.

There is increasing attention being paid to the conversion or valorization of solid wastes. Even in case of optimal source separation of waste aimed to material valorization, a residual amount remains and can be exploited for energy recovery through a thermochemical process. The research collected here in this volume includes four thermochemical processes: combustion, gasification, pyrolysis, and hydrothermal carbonization.

Combustion incinerates wastes, converting them into ash, flue gas, and heat. The ash—mostly the inorganic parts of the waste—is either solid lumps (slag) or particulates carried by the flue gas. To reduce the negative environmental impact, gaseous and particulate pollutants must be removed from the flue gases before they are emitted into the atmosphere. In some cases, the heat generated by incineration can be used to generate electric power. Even if the waste ultimately goes into a landfill, incineration radically reduces the amount of land required. Depending on the composition of the waste material, incinerators can reduce the solid mass of the original waste by as much as 80 to 85 percent; the volume of the original waste can be reduced up to 95 to 96 percent [1]. Recently it has been demonstrated in real scale that slag can be treated to recover metals and inert material for the building construction sector, thus reducing landfilling near to zero. Combustion may be particularly appropriate for the treatment of certain hazardous wastes, where the high temperature will destroy pathogens and toxins. Waste combustion tends to be popular in countries such as Japan where land is a scarce resource, while Denmark and Sweden have been using the energy generated from incineration for more than a century [2]. Worldwide, combustion is an option for the valorization of waste materials into energy. The modern age of waste combustion can be set since late in the 90s, when the European Union adopted more stringent limits for the emissions into the atmosphere.

Using high temperatures but without combustion, gasification converts carbonaceous materials into carbon monoxide, hydrogen, and carbon dioxide. The gasification method to produce energy has been in use for nearly two centuries. It was initially developed in the nineteenth century to produce gas for lighting and cooking from peat and coal. Since the 1920s, gasification has been used to produce synthetic chemicals, and during both world wars, the shortage of petroleum created an increased need for gasification-produced fuels [3]. Today, it is another option for valorizing solid waste. Controlled amounts of oxygen and/or steam are used in the process, and the resulting gas mixture is called syngas, which can be used as fuel. Chemical processing of the syngas may produce other synthetic fuels as well.

Using high temperatures and an external source of heat, pyrolysis decomposes organic material with very little oxidation involved. When pyrolysis is applied to organic substances, it produces gas and liquid products, while leaving a solid residue called char that is rich in carbon. The lignocellulose in crop wastes can provide feedstock for pyrolysis to create synthetic diesel fuel. Pyrolysis can also be used on plastic waste to produce fuel similar to diesel. In fact, many waste materials can be used as feedstock for pyrolysis. These include greenwaste, saw dust and waste wood, nut shells, straw, cotton trash, rice hulls, switch grass, poultry litter, cattle manure, waste paper, paper sludge, distillers grain, sewage sludge, and many more [4].

Hydrothermal carbonization is a relatively new variation of converting biomass into biofuel [5]. Moderate temperatures and pressures are used over an aqueous solution of biomass in a dilute acid for several hours. The resulting matter captures carbon for use as a solid fuel.

The transition from a fossil fuel-based economy to a more sustainable economy will not come quickly and without effort. It will require a solid foundation of ongoing scientific research. The articles included in this volume offer valuable building blocks to that foundation.

Elena Cristina Rada, PhD

REFERENCES

1. Ramboli Foundation. "Waste to Energy in Denmark," http://www.ramboll.com/about-us/ramboll-foundation. Retrieved 10-5-2015.
2. Kirkeby, Janus, Poul Erik Grohnheit, and Frits Møller Anderson. "Experiences with Waste Incineration for Energy Production in Denmark," Technical University of Denmark, 2014.
3. Santangelo, Steven, Philip Darcy, David Waage. "Waste to Energy at SUNY Cobleskill," Defense Technical Information Center, 2011.
4. Carpenter, Daniel, Tyler L. Westover, Stefan Czernika, and Whitney Jablonskia. Green Chemistry 16 (2014): 384–406
5. Titirici, Maria-Magdalena, Arne Thomas and Markus Antonietti. "Back in the Black: Hydrothermal Carbonization of Plant Material as an Efficient Chemical Process to Treat the CO2 Problem?" New Journal of Chemistry 31 (2007): 787–789.

Chapter 1, by Le Gleau and colleagues, deals with the treatment of acid gases present in fumes from meat and bones meal and sewage sludge co-incineration. A new implementation of acid gas treatment consists in catalytic ceramic filters, which are able to simultaneously capture particles, neutralise acid gases by dry basic sorbent injection and catalyse the reduction of nitrogen oxides by NH_3. This work aims at evaluating the efficiency of an existing system of industrial flue gas treatment, involving two different processing pathways installed in parallel. The results presented here consist in a measured material balance of the main species present in the fumes, followed by a discussion on the real processes and yields gotten under industrial conditions.

Fly ashes both from municipal solid waste incinerator (MSWI) and medical waste incinerator (MWI) are classified as hazardous materials because they contain high amounts of heavy metals. In present these contaminant ashes have become a major environmental problem. In Chapter 2, Pan and Xie determine the ability of these contaminating heavy metals to be incorporated into a glass-matrix and in various mineral phases after a high temperature melting process using a direct current plasma torch. After the melting process, the leaching characteristics of heavy metals in fly ash and vitrified slag were investigated using the toxicity characteristic leaching procedure (TCLP), and the products also were characterized by X-ray diffractometry (XRD) for crystal structure determination, and scanning electron microscopy (SEM) for microstructure/morphology observation. After vitrification, there were prominent changes in microstructures and crystalline phases between produced slags and raw ashes. TCLP results indicate the leaching level of heavy metals in slags decreases obviously and additives such as silica and liquid ceramic (LC) contribute to high effect on immobilization of heavy metals in a host glass matrix.

There is a growing recognition that conversion or valorization of wastes (municipal solid wastes (MSW) and agricultural wastes) is an environmentally responsible way of treating the increased volumes without occupying a large portion of land. Many companies are turning to gasification of MSW, biomass and mixtures of these to produce fuels and chemicals. In Chapter 3, Castaldi utilizes rapid heating rates (greater than or equal to 700°C min^{-1}) to gasify Clean Wood and two Refuse Derived Fuel (RDF) samples using thermogravimetric analysis coupled to gas chromatography (TGA/GC). Reaction atmospheres included Air, 5% O_2/95% Argon (Ar), 10% O_2/90% Ar, 100% Ar and steam and were used to produce gas evolution profiles for hydrocarbons ranging from H_2 to C_4H_{10}. While expected results were obtained using Air reaction atmospheres, some interesting results were observed using steam and Ar. Different concentration profiles and production rates of C_2H_6 compared to C_2H_4 and C_2H_2 enabled some understanding of the reaction sequence occurring during gasification under rapid heating conditions. Kinetic analysis showed pre-exponential factors of 8.00×10^{27}, 2.02×10^{29} and 3.71×10^{23} (sec^{-1} K1/2) for samples Clean Wood, RDF C (Industrial Solid Waste basis) and A (Municipal Solid Waste basis), respectively. Furthermore the apparent activation energy was determined to be 22, 71, and 185 (kJ mol^{-1}) for Clean Wood, RDF C and A respectively indicating that the Clean Wood is slightly more reactive than RDF C and more reactive than RDF A. This study also demonstrates good potential for H_2 production through gasification of RDF (from Industrial and Municipal Solid Wastes) in comparison to Clean Wood gasification. These RDF preparations were not specifically formulated with the intent of being used as gasification feedstocks. However, the present results show that gasification performance can be greatly improved by adjusting the feedstock formulation when employing RDF from Commercial, Industrial and Municipal Solid Wastes.

Recycling and utilization of waste is one of the key parameters of environmental issue. Chapter 4, by Tilahun and colleagues, explored the capability of supercritical waste gasification to convert the waste into marketable by-product. Adding catalysts or oxidants to supercritical waste gasifier can further reduce operating costs by creating self-sustaining reactions under mild conditions with even shorter residence times. The hydrogen produced by this process will be utilized for generating electricity using fuel cell technology. Besides, alkaline fuel cells appear to be an important technology in the future as they can operate at a high efficiency. Therefore, the combination of biomass gasification through supercritical water with alkaline fuel cells represents one of the most potential applications for highly efficient utilization of biomass. The main aim of the study is to recover energy from waste using alkaline fuel cell. With the different operation conditions 88.8 % of hydrogen and 45 % of carbon dioxide, maximum power density 9.24 W/cm^2 was obtained.

Chapter 5, by Tanigaki and colleagues, evaluates municipal solid waste co-gasification technology and a new solid waste management scheme, which can minimize final landfill amounts and maximize material recycled from waste. Waste is processed with incombustible residues and an incineration bottom ash discharged from existent conventional incinerators, using a gasification and melting technology. The co-gasification system produced high quality slag with few harmful heavy metals, which was recycled completely without requiring any further post-treatment. As a consequence, the co-gasification system had an economical advantage over other systems. Sensitivity analyses of landfill cost were also conducted. The higher the landfill costs, the greater the advantage of the co-gasification system has. The co-gasification was beneficial for landfill cost in the range of 80 Euro per ton or more. These results indicate that co-gasification of bottom ash and incombustibles with municipal solid waste contributes to minimizing the final landfill amount and has great possibilities maximizing material recovery and energy recovery from waste.

Organic and inorganic contaminants in sewage sludge may cause their presence also in the by-products formed during gasification processes. Thus, Chapter 6, by Werle and Dudziak, presents multidirectional chemical instrumental activation analyses of dried sewage sludge as well as both solid (ash, char coal) and liquid (tar) by-products formed during sewage gasification in a fixed bed reactor which was carried out to assess the extent of that phenomenon. Significant differences were observed in the type of contaminants present in the solid and liquid by-products from the dried sewage sludge gasification. Except for heavy metals, the characteristics of the contaminants in the by-products, irrespective of their form (solid and liquid), were different from those initially determined in the sewage sludge. It has been found that gasification promotes the migration of certain valuable inorganic compounds from sewage sludge into solid by-products which might be recovered. On the other hand, the liquid by-products resulting from sewage sludge gasification require a separate process for their treatment or disposal due to their considerable loading with toxic and hazardous organic compounds (phenols and their derivatives).

In Chapter 7, Zhao and colleagues investigate the pyrolysis of waste plastics separated from municipal solid wastes (MSW) and pyrolysis of whole combustibles in MSW together to compare their products and emissions with a purpose to improve the pyrolysis process for combustibles in MSW. The pyrolysis experiments were carried out with two substrates, namely the upper siftings from excavated aged MSW in landfill cell and waste plastics separated from the upper siftings. The upper siftings were indeed combustibles and their composition was also investigated. The characteristics of pyrolysis products were studied with special attention paid to mass distribution among gas, liquid and char products, composition of gas products and liquid products, quality of oil products and pollutants in gas products. Chars from the two pyrolysis processes were also investigated to check their possible applications and their leaching characteristics. The results obtained based on the tests carried out at lab scale proved that the pyrolysis process carried out for the separated waste plastics was preferred to pyrolysis process for the whole upper siftings. Those information can be very useful for the design of a pyrolysis process for combustibles or waste plastics from MSW.

The amount of plastic waste is growing every year and with that comes an environmental concern regarding this problem. The authors of chapter 8 explore pyrolysis as a tertiary recycling process as a solution. Pyrolysis can be thermal or catalytical and can be performed under different experimental conditions. These conditions affect the type and amount of product obtained. With the pyrolysis process, products can be obtained with high added value, such as fuel oils and feedstock for new products. Zeolites can be used as catalysts in catalytic pyrolysis and influence the final products obtained.

A novel strategy of waste recycling of polypropylene plastics (PP) bags for generation of commercially viable byproducts using nanoforms of nickel as catalyst is presented in Chapter 9, by Mishra and colleagues. After pyrolysis of waste PP bags (>20 μm) under continuous argon flow, 90% conversion efficiency to high petroleum oil was observed at 550°C. To assess the physicochemical attributes of formed oil, flash point, pour point, viscosity, specific gravity, heating value, and density were also measured and found to be very close to ideal values of commercial fuel oil. Moreover, GC-MS was used to resolve the range of trace mass hydrocarbon present in the liquefied hydrocarbon. This robust recycling system can be exploited as economical technique to solve the nuisance of waste plastic hazardous to ecosystem.

Chapter 10, by Sun and colleagues, describes a kinetic study of the decomposition of waste printed circuit boards (WPCB) under conventional and microwave-induced pyrolysis conditions. We discuss the heating rates and the influence of the pyrolysis on the thermal decomposition kinetics of WPCB. We find that the thermal degradation of WPCB in a controlled conventional thermogravimetric analyzer (TGA) occurred in the temperature range of 300 °C–600 °C, where the main pyrolysis of organic matter takes place along with an expulsion of volumetric volatiles. The corresponding activation energy is decreased from 267 kJ/mol to 168 kJ/mol with increased heating rates from 20 °C/min to 50 °C/min. Similarly, the process of microwave-induced pyrolysis of WPCB material manifests in only one stage, judging by experiments with a microwave power of 700 W. Here, the activation energy is determined to be only 49 kJ/mol, much lower than that found in a conventional TGA subject to a similar heating rate. The low activation energy found in microwave-induced pyrolysis suggests that the adoption of microwave technology for the disposal of WPCB material and even for waste electronic and electrical equipment (WEEE) could be an attractive option.

In Chapter 11, by Sarkar and colleagues, newspaper waste was pyrolysed in a 50 mm diameter and 640 mm long reactor placed in a packed bed pyrolyser from 573 K to 1173 K in nitrogen atmosphere to obtain char and pyro-oil. The newspaper sample was also pyrolysed in a thermogravimetric analyser (TGA) under the same experimental conditions. The pyrolysis rate of newspaper was observed to decelerate above 673 K. A deactivation model has been attempted to explain this behaviour. The parameters of kinetic model of the reactions have been determined in the temperature range under study. The kinetic rate constants of volatile and char have been determined in the temperature range under study. The activation energies 25.69 KJ/mol, 27.73 KJ/mol, 20.73 KJ/mol and preexponential factors 7.69 min^{-1}, 8.09 min^{-1}, 0.853 min^{-1} of all products (solid reactant, volatile, and char) have been determined, respectively. A deactivation model for pyrolysis of newspaper has been developed under the present study. The char and pyro-oil obtained at different pyrolysis temperatures have been characterized. The FT-IR analyses of pyro-oil have been done. The higher heating values of both pyro-products have been determined.

Chapter 12, by Lam and colleagues, presents an extensive review of the scientific literature associated with various microwave pyrolysis applications in waste to energy engineering. It was established that microwave-heated pyrolysis processes offer a number of advantages over other processes that use traditional thermal heat sources. In particular, microwave-heated processes show a distinct advantage in providing rapid and energy-efficient heating compared to conventional technologies, and thus facilitating increased production rates. It can also be established that the pyrolysis process offers an exciting way to recover both the energetic and chemical value of the waste materials by generating potentially useful pyrolysis products suitable for future reuse. Furthermore, this review has revealed good performance of the microwave pyrolysis process when compared to other more conventional methods of operation, indicating that it shows exceptional promise as a means for energy recovery from waste materials. Nonetheless, it was revealed that many important characteristics of the microwave pyrolysis process have yet to be raised or fully investigated. In addition, limited information is available concerning the characteristics of the microwave pyrolysis of waste materials. It was thus concluded that more work is needed to extend existing understanding of these aspects in order to develop improvements to the process to transform it into a commercially viable route to recover energy from waste materials in an environmentally sustainable manner.

In Korea, municipal solid waste (MSW) treatment is conducted by converting wastes into energy resources using the mechanical-biological treatment (MBT). The small size MSW to be separated from raw MSW by mechanical treatment (MT) is generally treated by biological treatment that consists of high composition of food residue and paper and so forth. In Chapter 13, Kim and colleagues apply the hydrothermal treatment to treat the surrogate MT residue composed of paper and/or kimchi. It was shown that the hydrothermal treatment increased the calorific value of the surrogate MT residue due to increasing fixed carbon content and decreasing oxygen content and enhanced the dehydration and drying performances of kimchi. Comparing the results of paper and kimchi samples, the calorific value of the treated product from paper was increased more effectively due to its high content of cellulose. Furthermore, the change of the calorific value before and after the hydrothermal treatment of the mixture of paper and kimchi can be well predicted by this change of paper and kimchi only. The hydrothermal treatment can be expected to effectively convert high moisture MT residue into a uniform solid fuel.

PART I

COMBUSTION

CHAPTER 1

Comparative Study of Flue Gas Dry Desulphurization and SCR Systems in an Industrial Hazardous Waste Incinerator

F. LE GLEAU, S. CAILLAT, E. PERDRIX, L. GASNOT, D. GAMBIER, AND J-F. PAUWELS

1.1 INTRODUCTION

The studied plant co-incinerates two hazardous wastes.

The first one, meat and bone meal (MBM), is unmarketable by-product and waste from slaughterhouses. Indeed, since 1994 and the bovine spongiform encephalopathy crisis, the use of MBM for animal feeding is forbidden in Europe. France produces 850 000 t per year of MBM (Conesa et al., 2005). As the low heating value (LHV) of MBM is about 14.5 to 30 $MJ.kg^{-1}$ (Senneca, 2008; Skodras et al., 2007), that is to say equivalent or higher than wood, most of it is burned with coal in cement works, the remaining being used for electricity or heat production.

Le Gleau F, Caillat S, Perdrix E, Gasnot L, Gambier D, and Pauwels J-F. "Comparative Study of Flue Gas Dry Desulphurisation and SCR Systems in an Industrial Hazardous Waste Incinerator" Proceedings Venice 2010, Third International Symposium on Energy from Biomass and Waste, *Venice, Italy; 8-11 November 2010 © CISA, Environmental Sanitary Engineering Centre, Italy (2010). Used with permission from the publisher.*

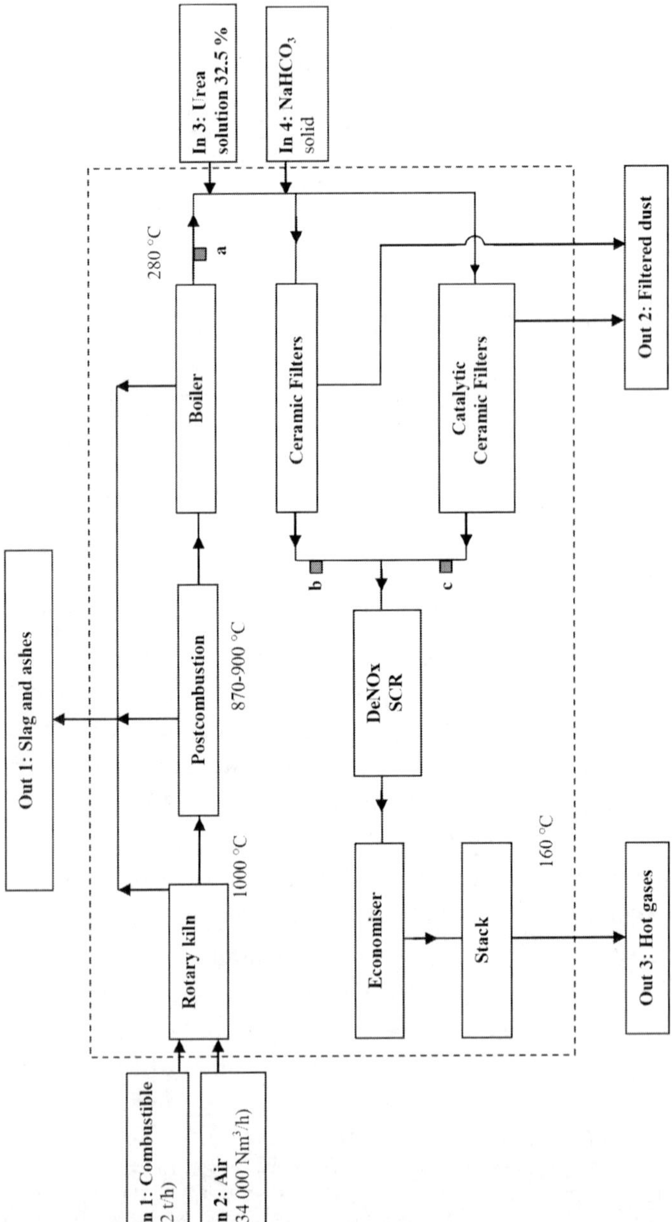

FIGURE 1: Synoptic of the MBM and sewage sludges co-incineration plant.

The second waste is sewage sludge. About 850 000 t of dry matter are produced in France yearly, mainly used in crop fertilising, but also dumped or incinerated. The LHV of dry sewage sludge can reach 20 $MJ.kg^{-1}$ (Murakami et al., 2009, Leckner et al., 2004).

The incineration of these materials generates non-combustible ashes. Deyder et al. (2005) showed that the amount of these ashes can equal a quarter of the original mass for MBM. Moreover, acids gases are emited during incineration (mainly SO_2, HCl and NOx), their treatment generates wastes.

In this study, the efficiency of a waste treatment and its associated flue gas cleaning process was investigated by doing a mass balance of the main species present in the fumes of an industrial waste incineration plant.

1.2 INSTALLATION AND MEASUREMENT

1.2.1 STUDIED INSTALLATION

The studied installation (Figure 1) is designed to incinerate 2 t/h of wastes, which generates until 40 000 Nm^3/h of fumes. Wastes are incinerated in a rotary kiln (Ndiaye et al., 2009). The hot gas passes the post-combustion zone where volatile matter is burnt and then through the boiler for heat recovery.

The first flue gas cleaning process installed on the plant consisted in a neutralisation of acid gases by duct injection of dry sodium bicarbonate with subsequent filtration on ceramic filters, followed by a Selective Catalytic Reduction (SCR) of NOx by urea. The gas cleaning system of the plant was then upgraded by adding in parallel a second flue gas cleaning process, more innovative, which consists in the simultaneous duct injection of aqueous urea and dry sodium bicarbonate, followed by a combined filtration and SCR by CERCAT ceramic catalytic filters. They consist in ceramic filters impregnated of V_2O_5 catalyst. This innovative system enables a space saving. The two processing pathways treat the same fumes simultaneously and allowed therefore their comparison. The Figure 1 is a synoptic of the plant: black squares indicate the gas sampling locations.

In 1 to In 4 and Out 1 to Out 3 are respectively the 4 inlet and the 3 outlet considered in this study.

1.2.2 ANALYSIS METHODS

1.2.2.1 GAS ANALYSES

The O_2, CO, CO_2, NOx and SO_2 concentrations were analysed on-line for 24 h by a Horiba PG250 analyser. The gas is sampled through a heated probe, filtered, dried and sent to the analyser. O_2 is measured by paramagnetic method, while CO, CO_2 and SO_2 are measured by infrared spectrometry and NOx by chemiluminescence. H_2O and HCl have been sampled by bubblers during 3 sampling phases of 3 h each. The percentage of H_2O in flue gas was determined by weighting and the concentration of HCl was measured by ion chromatography (IC).

These chemical species were measured at the outlet of the boiler (Point a on Figure 1), in order to get the gas composition before any treatment and at the outlet of each filter (Points b and c on Figure 1) in order to compare their efficiencies.

1.2.2.2 SOLIDS ANALYSES

Solids (In 1, Out 1 and Out 2 on Figure 1) were sampled then stocked in the dark at room temperature. The analyses of C, H, N, S, Cl were done following international or French standards: ISO/TS 12902 for C and H, NF M 03-018 for N, NF EN 14 582 for S and Cl.

The samples were analysed by X-Ray Fluorescence (XRF) for a whole elemental analysis. A sample mass of 1 g was pelletized with lithium tetraborate. The XRF spectrometer (Bruker S4 Pioneer) uses a Rhodium X-ray source and the detector is a gas flow proportional counter.

In order to measure hydrosoluble ions (Cl^-, NO_3^-, SO_4^{2-} and Na^+) IC analyses were done on leachate of solids. A mass of 2 g of sample was mixed with 20 mL of demineralised water for 24 h, and the supernatant was analysed by a DX-120 Dionex, the anions on a column Ion Pac AS4A and the cations on a column Ion Pac CS16.

1.2.3 FLUE GAS TREATMENT

As regards acid gas abatement, the two flue gas treatment pathways are based on the same chemical process: SO_2 and HCl react with Na_2CO_3 to form salts which are filtered, while NO is reduced into N_2 by SCR with NH_3. More precisely, the $NaHCO_3$ injected in the duct decomposed itself into Na_2CO_3, (equation 1) and, on one hand, reacts with SO_2 to form Na_2SO_3 which is oxidised in Na_2SO_4, following equation 2 and 3 (Wu et al., 2004). On the other hand, Na_2CO_3 also react with HCl to form NaCl, following equations 4 (Duo et al., 1996).

$$2NaHCO_3 \rightarrow Na_2CO_3 + H_2O + CO_2 \qquad (1)$$

$$Na_2CO_3 + SO_2 \rightarrow Na_2SO_3 + CO_2 \qquad (2)$$

$$Na_2SO_3 + 1/2O_2 \rightarrow Na_2SO_4 \qquad (3)$$

$$Na_2CO_3 + 2HCl \rightarrow 2NaCl + H_2O + CO_2 \qquad (4)$$

For the NO treatment, the urea solution injected in the pipe dries and decomposes itself into ammonia in two steps following the equations 5 and 6 (Zanoelo et al., 2009). NH_3 reacts with NO to form N_2 and water following equation 7 (Busca et al., 1998). Because the non-catalytic reduction of NO cannot occur at temperatures lower than 300 °C (Rota & Zanoelo, 2003) the reaction needs to be catalysed. Here a catalyst based on vanadium oxide is used both in the SCR reactor and in the catalytic ceramic filters.

$$CO(NH_2)_2 \rightarrow NH_3 + HNCO \tag{5}$$

$$HNCO + H_2O \rightarrow NH_3 + CO_2 \tag{6}$$

$$4NO + 4NH_3 + O_2 \rightarrow 4N_2 + 6H_2O \tag{7}$$

1.3 RESULTS AND DISCUSSION

1.3.1 MAIN SPECIES MASS BALANCE

In order to study the efficiency of the waste treatment plant a material balance of the main species was made. Each flue (in & out) is presented in Figure 1 and the methods to determine their concentrations are presented in Table 1.

The first inlet flue (In 1) is the fuel composed by a dried mixture of MBM (75%) and sewage sludge (25 %). The second one (In 2) is the combustion air, which comprises the air flow used for drying the combustible and recycled to avoid odour emissions. It therefore contains water. The third and fourth inlet flues (In 3 and In 4) are the injection of urea solution at 32.5 % (mass) and the $NaHCO_3$ solid used for the gas cleaning.

The first outlet flue (Out 1) is slag and ashes mixed; the second one (Out 2) is the sum of the filtration residues from the catalytic and the non-catalytic filters. The third outlet flue (Out 3) is the stack gases; the contribution of the particulate matter present in the fumes was found negligible (less than 2 g/h).

The main elements present in the system are, C, H, O, N, S, Cl and Na. Figure 2 presents the flow rates of the elements analysed in each flow for a normalised fuel flow of 1 t/h.

	Nature	*State*	*Analyses*						
			C	*H*	*O*	*N*	*S*	*Cl*	*Na*
In 1	MBM/Sewage sludge	Solid	ISO/TS 12902		calculated	NF M 03-018	NF EN 14 582		XRF
In 2	Air	Gas	Theoretical composition						
In 3	Urea solution	Liquid	Theoretical composition						
In 4	Sodium Hydrogenocarbonate	Solid	Theoretical composition						
Out 1	Slag and ashes	Solid	ISO/TS 12902		calculated	NF M 03-018	XRF, IC		
Out 2	Filtered dust	Solid	ISO/TS 12902		calculated	NF M 03-018	XRF, IC		
Out 3	Stack gases	Gas	Horiba					bubbling-IC	-

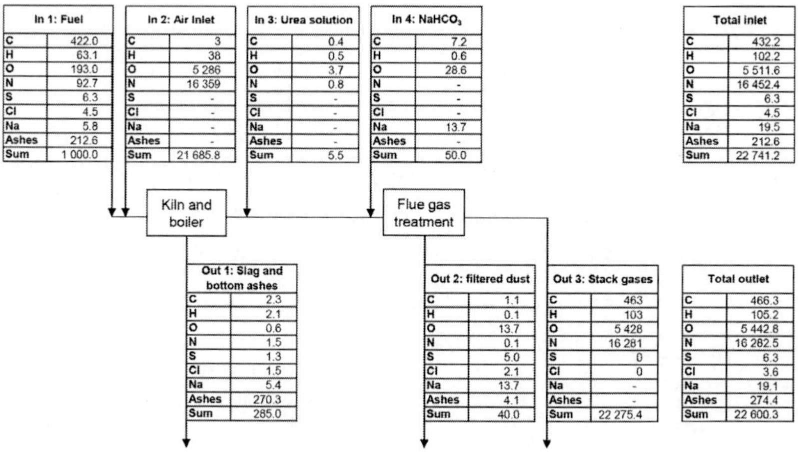

FIGURE 2: Flow rates in kg.h^{-1} of the elements analysed in each flue of the system for a normalised fuel flow of 1 t/h.

TABLE 1: Methods used to determine the element concentrations in each flow of the system.

	Nature	State	Analyses						
			C	H	O	N	S	Cl	Na
In 1	MBM/Sewage sludge	Solid	ISO/TS 12902		calculated	NF M 03-018	NF EN 14 582		XRF
In 2	Air	Gas				Theoretical composition			
In 3	Urea solution	Liquid				Theoretical composition			
In 4	Sodium Hydrogeno-carbonate	Solid				Theoretical composition			
Out 1	Slag and ashes	Solid	ISO/TS 12902		calculated	NF M 03-018		XRF, IC	
Out 2	Filtered dust	Solid	ISO/TS 12902		calculated	NF M 03-018		XRF, IC	
Out 3	Stack gases	Gas			Horiba			bubbling IC	–

The mass balance was validated by calculating the maximum error for each element X defined by the equation 8:

$$E_X = \frac{\sum inlet_x - \sum outlet_x}{Min(\sum inlet_x; \sum outlet_x)} \tag{8}$$

All the errors calculated are lower than 10 % except for Cl and ashes, for which the error is respectively 23.7 % and -29.1 % (Table 2). This could be because of an underestimation of the Cl and an overestimation of the ashes from the mixture of MBM and sewage sludges coming from the high heterogeneity of these wastes.

Figure 2 shows that the initial sulphur of the fuel distributes on one hand in the slag and bottom ashes (21 %) and on the other hand in the filtered dust stack gas (79 %). For one ton of fuel, the studied plant generates 285 kg of slag and bottom ashes, which are valorisable as road fill, and 40 kg of filtered dust containing the Na_2SO_4 and NaCl from the air cleaning process which are not valorised. This waste treatment divides by 3 the mass of wastes, produces 28 % of secondary wastes reusable as road fill and 4 % of residual wastes which have to be dumped.

TABLE 2: Calculated error for each element.

Error	C	H	O	N	S	Cl	Na	Ashes	Sum
%	-7.9	-3.0	1.3	1.0	-0.8	23.7	1.8	-29.1	0.6

With an estimated LHV of MBM and sewage sludges mixed of 23 MJ/kg, the power provided by the waste combustion, is about 6.2 MW/t. Approximately 70 % of energy is recovered in the boiler. The Table 3 presents average characteristics of municipal waste incineration plant (Menard, 2003) for comparison with the MBM incineration plant. The ashes and filtered dust emitted are at the same order of magnitude, but the power produced is almost 3 times higher here.

TABLE 3: Comparison of different incineration plants.

Wastes	Ashes (kg/t)	Filtered dust (kg/t)	LHV (MJ/kg)
MBM 75 % + sewage sludges 25%	285	40	23.0
Municipal waste (Menard, 2003)	250-300	25 - 50	7.8

1.3.2 EFFICIENCY OF THE GAS CLEANING

Removal efficiency in the gas cleaning system is defined by the equation 9:

$$\eta_X = \frac{[X]_a - [X]_{b/c}}{[X]_a} \times 100 \tag{9}$$

where $[X]_i$ is the concentration in $mg.Nm^{-3}$ of the compound X (SO_2, HCl or NO) at the Point i.

Efficiency is calculated assuming that the modification of the total volumic flow rate due to the addition of urea and $NaHCO_3$ is negligible.

The SO_2 and HCl concentrations measured (Table 4) at the outlet of both filters were below the quantification limits, indicating a treatment efficiency of more than 95 % for the SO_2 and more than 97 % for HCl. The SO_2 and HCl abatement are equally efficient on both filters, and the catalytic filters do not appear to be poisoned by acid gas or particulate matter since their installation in January 2008.

Concerning the SCR, the quantity of urea injected is set to reach a given NOx concentration, at the outlet of the catalytic filter, lower than the regulatory emission limit of 200 mg/Nm^3 at 11 % O_2 on dry fumes (European directive 2000/76/EC). Considering the quantity of ammonia injected, the NOx concentration was lower than if abatement was only due to the reaction 7.

TABLE 4: SO_2, HCl and NOx concentration at 11 % of O_2 on dry fumes at the outlet of the boiler and the outlet of each filters and the efficiency of each filter.

Acid gas	Boiler outlet	Non catalytic filter outlet	Catalytic filter outlet	SCR Reactor outlet	Non catalytic filter efficiency	Catalytic filter efficiency
	mg/Nm^3	mg/Nm^3	mg/Nm^3	mg/Nm^3	%	%
SO_2	726	< 30	< 30	< 30	> 95	> 95
HCl	256	< 5	< 5	< 5	> 97	> 97
NOx	460	360	185	~ 95	22	60

Removal efficiency of 22 % for NOx in the non-catalytic filter was also observed. As the operating temperature is too low for a SNCR, we assume that another process does occur: either a reaction with Na_2CO_3 with formation of $NaNO_3$ as described as thermodynamically possible by Verdone & De Filippis (2004) or a NO adsorbtion on the fly ash or a SCR catalysed by metal oxides from the fly ashes. Further investigations are needed to clarify this reduction.

At the filters outlet, the fumes come together and pass through a SCR reactor to reach a NO concentration lower than 200 mg/Nm^3 at 11 % O_2 on dry fumes.

1.4 CONCLUSION

A MBM and sewage sludge co-incineration plant gas cleaning system was investigated, two ceramic filtrations systems, catalytic and non catalytic, were compared by a mass balance of main chemical species (C, H, O, N, S, Cl & Na). A good efficiency of the gas cleaning was observed despite the high quantity of gas emitted due to the important concentrations of S, Cl and mostly N in the wastes mixture. NOx abatement was also observed in the non-catalytic filter, this could be a reaction of NOx with the sodium carbonate to form $NaNO_3$, the fly ashes may also adsorb NOx, or the metal oxides they contain may catalyse the reaction of NO with NH_3. Further investigations are necessary to a better understanding of these phenomena.

REFERENCES

1. Busca G., Lietti L., Ramis G. and Berti F. (1998). Chemical and mechanistic aspects of the selective catalytic reduction of NOx by ammonia over oxide catalysts: A review, Appl. Catal., B Environ., 18, 1-36.
2. Conesa J.A., Fullana A. and Font R. (2005). Dioxine production during the thermal treatment of meat and bone meal residues, Chemosphere, 59, 85-90.
3. Deyder E., Richard G., Sarda S., Sharrock P. (2005). Physical and chemical characterisation of crude meat and bone meal combustion residue: "waste or raw material?", Journal of Hazardous Materials, B121, 141-148.
4. Directive 2000/76/EC of the European Parliament and of the Council of 4 December 2000 on the incineration of waste, OJ L 332 of 28.12.2000
5. Duo W., Kirkby N.F., Seville J.P.K., Kiel J.H.A., Bos A. and Den Uil H. (1996). Kinetics of reaction with calcium and sodium sorbents for IGCC fuel gas cleaning, chem. Eng. Sci., 51, 11, 2541-2546.
6. Leckner B., Amand L.-E., Lücke K. and Werther J. (2004). Gaseous emissions from co-combustion of sewage sludge and coal/wood in a fluidized bed, Fuel, 83, 477-486.
7. Menard Y. (2003). Modelisation de l'incinération sur grille d'ordures ménagères et approche thermodynamique du comportement des métaux lourds (Mathematical Modelling of Municipal Solid Waste Incineration and Thermodynamic Study of the

Behaviour of Heavy Metals, in French), PhD thesis, Institut National Polytechnique de Lorraine.
8. Murakami T., Suzuki Y., Nagasawa H., Yamamoto T., Koseki T., Hirose H. and Okamoto S. (2009). Combustion characteristics of sewage sludge in an incineration plant for energy recovery, Fuel Process. Technol., 90, 778-783.
9. Ndiaye L.G., Caillat S., Chinnayya A., Gambier D. and Baudoin B. (2010). Application of the dynamic model of Saeman to an industrial rotary kiln incinerator: numerical and experimental results, Waste Management, Volume 30, Issue 7, 1188-1195.
10. Rota R. and Zanoelo E.F. (2003). Influence of oxygenated additives on the NOxOUT process efficiency, Fuel, 82, 765-770.
11. Senneca O. (2008). Characterisation of meat and bone mill for coal co-firing, Fuel, 87, 3262-3270.
12. Skodras G., Grammelis P. and Basinas P. (2007). Pyrolysis and combustion behaviour of coal-MBM blends, Bioresour. Technol., 98, 1-8.
13. Verdone N. and De Filippis P. (2004). Thermodynamic behaviour of sodium and calcium based sorbents in the emission control of waste incinerators, Chemosphere, 54, 975-985.
14. Wu C., Khang S-J., Keener T.C. and Lee S-K. (2004). A model for dry sodium bicarbonate duct injection flue gas desulfurization, Advances in Environmental Research, 8, 655-666.
15. Zanoelo E.F. (2009). A lumped model for thermal decomposition of urea. Uncertainties analysis and selective non-catalytic reduction of NO. Chemical Engineering Science, Chem. Eng. Sci., 64, 1075-1084

CHAPTER 2

Characteristics of Melting Incinerator Ashes Using a Direct Current Plasma Torch

XINCHAO PAN AND ZHENGMIAO XIE

2.1 INTRODUCTION

Over the past two decades, incineration has been increasingly applied for treating municipal solid waste (MSW). The dominating purpose of burning MSW is to cut down the volume and mass of MSW, because of the increasing difficulty of finding suitable sites for controlled and uncontrolled landfill waste disposal operations [1,2]. The capacity of incinerating MSW was about 16.4 Mt, up to 16% of all MSW in 2011 in China [3]. Recently, incineration has also been the most widespread adopted technology for the disposal of medical waste (MW) since the nationwide outbreak of severe acute respiratory syndrome (SARS) in 2003 in China. Incinerating MW can not only neutralize its infectivity, which is the most hazardous MW property, but also sharply reduce its volume [4,5].

Characteristics of Melting Incinerator Ashes Using a Direct Current Plasma Torch. © *Pan X and Xie Z.* Journal of Environmental and Analytical Toxicology ***4****,212 (2014). doi: 10.4172/2161-0525.1000212.*

However, during incinerating MSW and MW, a great quantity of fly ashes from the air pollution control (APC) systems, which were set up for wiping off hazardous materials in the flue gas, were discharged. These APC fly ashes were classified as hazardous wastes owing to containing significant amounts of toxic materials such as heavy metals and persistent organic pollutants (POPs), e.g. dioxins and furans [6-9], and the fly ash from medical waste incinerator (MWI) was much more deleterious than that of municipal solid waste incinerator (MSWI) [8,9].

With growing public concerns and rigorous regulatory requirements, how to safely handle the ash is gaining more and more attention by the scientific community and by the general society. It is well known that inappropriate treatment and final disposal of the ash can induce adverse impacts on both public health and the environment [10]. Therefore, developing a safe and reliable immobilization technology, to transform the ash into a stable form, is very necessary. Many alternative methods for hazardous fly ash treatment have been suggested and developed. One of these methods is the melting technology which reduces the volume, yields the glassy leaching-resistant slag, and destroys toxic organic compounds effectively [11-14].

Among several melting methods, plasma melting technology has attracted increasing interest for hazardous waste treatment. Compared with a fuel type melting furnace, a thermal plasma system has the advantages of high temperature and high energy density, which allows fast heat transfer at the reactor boundaries and correspondingly shorter treatment time. In the past decade, thermal plasma technology has been extensively used for the treatment of various hazardous wastes [13-20].

In this study a direct current (DC) plasma appliance has been developed for the vitrification of fly ashes from both MWI and MSWI. The transformation of the mineralogical species of fly ashes and the leachability of major heavy metals during melting process were investigated, in order to ascertain the mechanism of fixing metal in molten slag. Also the density and the microstructure of original ash and vitrified slag were surveyed.

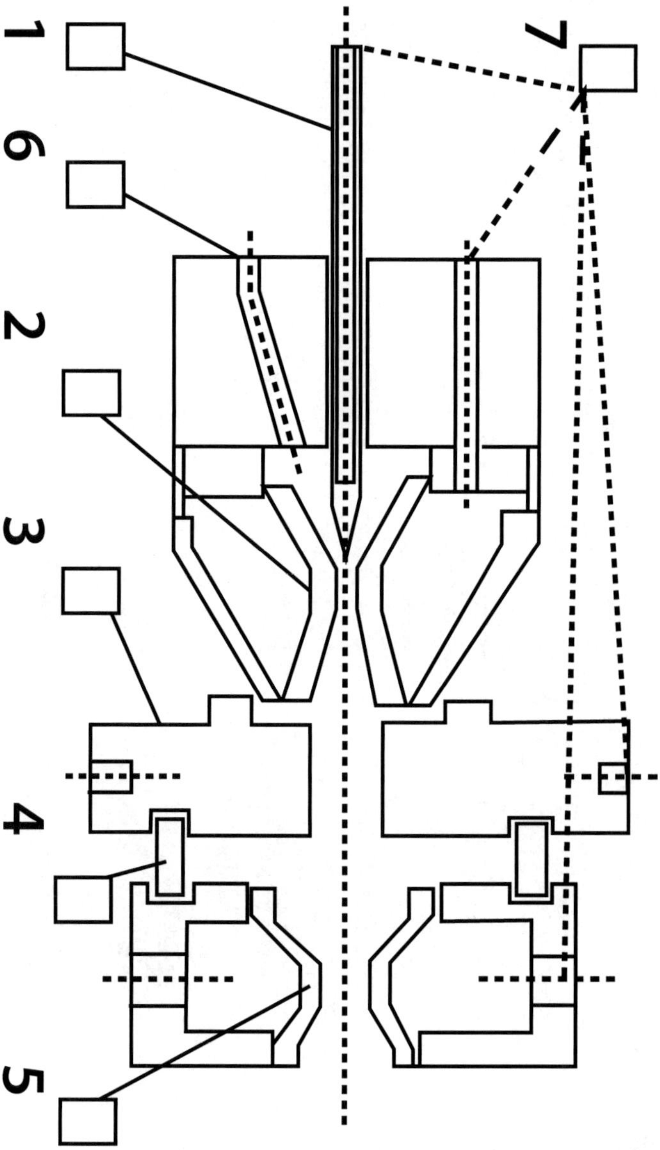

FIGURE 1: Schematic diagram of experimental set-up 1-Cathode, 2-Anode I, 3-Linked Part, 4-Insulated Ring, 5-Anode II, 6-Gas Entrance, 7-Cooling Water.

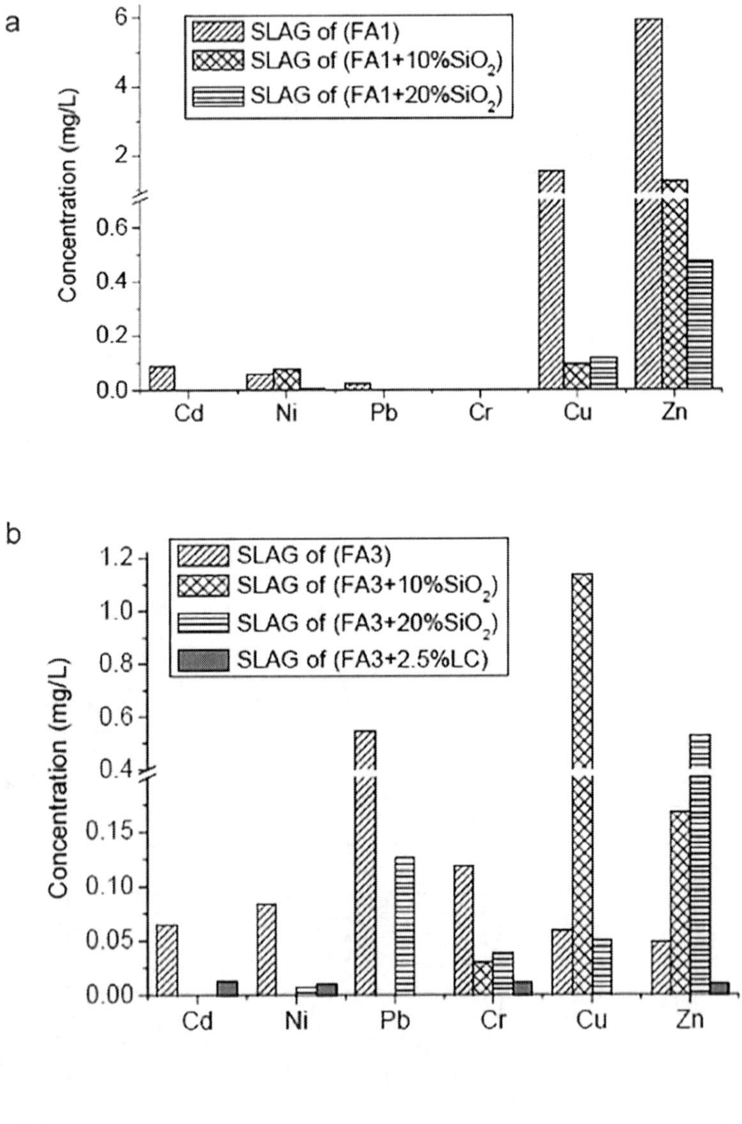

FIGURE 2: Leaching characteristics of heavy metals in slags (a) Leaching behavior of molten slags of blended ash (FA1 + SiO_2) (b) Leaching behavior of molten slags of blended ash (FA3+ SiO_2/LC)

2.2 EXPERIMENTAL METHODS

2.2.1 THERMAL PLASMA SYSTEM

The plasma torch used in this experiment consists of four major parts: cathode, first anode, linked part and second anode, as shown in Figure 1. Compared with conventional thermal plasma torches, this torch has a special design with two nozzles shaped copper anodes set at different axial distances from the cathode tip [21,22]. This configuration can not only extend jet length but also enhance the arc stability. In this work, argon was used as working gas at a flow rate varied from 12 to 14 L/min. The double arcs plasma torch was operated in direct current mode with typically 20-30 V/100 A for the first arc and 50-60 V/100 A for the second arc. The temperature of the plasma jet near the torch exit is over 11000K, and the heat flux of the plasma jet is around 65 kW/m^2 at 14cm downstream from the exit [20-24].

A plasma melting furnace based on this plasma torch has been developed in our lab. A crucible filled with fly ash was vitrified by atmospheric thermal plasma jet. The crucible had a capacity of 30-60 g of fly ash, which could be completely vitrified within about 15 min. using this plasma melting furnace, prior study on treating MWI fly ash has shown satisfactory results [11].

2.2.2 FLY ASH

Three representative fly ash samples were used in this investigation, named FA1, FA2 and FA3, respectively. Both FA1 and FA2 were collected from fabric filters in air pollution cleaning devices installed in MSWI, but the former was obtained from a grate type incinerator, then the latter was from a circulated fluidized bed incinerator. FA3 was sampled from a medium-scale MWI with handling capacity of 10 tons/d, equipped with a simplified stoker furnace. These three incinerators are all located in Zhejiang province in Southeast China.

TABLE 1: Elemental composition of fly ashes (wt%).

Element	FA1	FA2	FA3
C	8.69	25.35	13.54
O	30.80	33.15	23.90
Na	2.26	1.29	1.75
Mg	2.61	1.23	3.01
Al	3.59	8.83	4.28
Si	7.50	13.87	6.17
P	1.12	0.73	0.63
S	3.58	0.99	1.51
Cl	11.96	1.26	17.23
K	2.79	1.44	1.50
Ca	22.95	7.89	21.66
Fe	2.15	3.24	0.56
LOI	7.17	10.3	18.2
	heavy metals concentration (mg/kg)		
Pb	2943.5 (713.7)*	1964.2 (70.1)	1237.4 (214.1)
Cd	142.2 (15.0)	68.4 (14.8)	74.7 (13.3)
Cr	74.3 (21.4)	93.7 (7.7)	117.3 (34.3)
Zn	9743.8 (1511.4)	7512.5 (221.6)	8053.4 (780.2)
Cu	465.2 (43.2)	542.7 (75.9)	227.7 (21.1)
Ni	74.9 (34.9)	69.2 (33.7)	55.9 (31.2)

LOI - loss of ignition
**Values in parentheses are standard deviation of means of triplicate*

The elemental composition of samples tested by X-ray Energy Disperse Spectroscopy (EDS) (GENENIS 4000, EDAX Inc. USA) is shown in Table 1. The elementary composition of these three samples was very similar, and their primary elements were carbon, oxygen, silicon, chlorine and calcium. Nevertheless, the loss of ignition (LOI) was far different, and LOI of FA3 was as high as 18.2%, which was much higher than that of FA1 and FA2. That was to say the proportion of organic components in MWI ash was much higher, compared with MSWI ash, which went against melting process.

The heavy metal in ash samples was extracted using mingled acid solution then tested by Atomic Absorption Spectrophotometer (AAS) (SOLAAR 969, Thermo Inc., America). The concentrations are also listed in Table 1. As can be seen, the amount of Zn and Pb in all samples far exceeded that of Cd, Cr, Cu and Ni, and Zn was especially so.

TABLE 2: The leaching concentrations of heavy metals in fly ashes and vitrified slags (mg/L).

	Pb	Cd	Cr	Zn		Cu	Ni
FA1	0.3254 (0.070)*	0.1323 (0.0254)	0.2464 (0.0735)	0.1422 (0.0077)	0.0483	(0.0028)	0.2293 (0.0585)
FA2	0.2868 (0.1005)	0.3033 (0.0291)	0.4725 (0.2133)	4.7465 (1.230)	0.2859	(0.1407)	0.2851 (0.0753)
FA3	0.3106 (0.0786)	0.1361 (0.0195)	0.2491 (0.0128)	0.0758 (0.0034)	0.1109	(0.0136)	0.1973 (0.0141)
S1	0.0243 (0.0151)	0.0865 (0.0044)	BDL	5.9382 (0.011)	1.533	(0.026)	0.0504 (0.0236)
S2	BDL	BDL	BDL	0.0843 (0.0098)	0.2972	(0.0221)	0.308 (0.0258)
S3	0.4946 (0.1657)	0.0559 (0.0397)	0.1067 (0.0622)	0.0537 (0.0238)	0.0544	(0.0236)	0.0763 (0.0388)
China limit	3	0.3	10	50	50	10	

BDL-below detection limit
**Values in parentheses are standard deviation of means of triplicate*

2.3 RESULTS AND DISCUSSION

2.3.1 METAL LEACHING TEST

Fly ash and molten slag: The leaching capability of heavy metals was evaluated by the toxicity characteristic leaching procedure method (TCLP, USEPA method 1311). Acetic acid solution (pH 2.88 ± 0.05) was used as the leaching liquid.

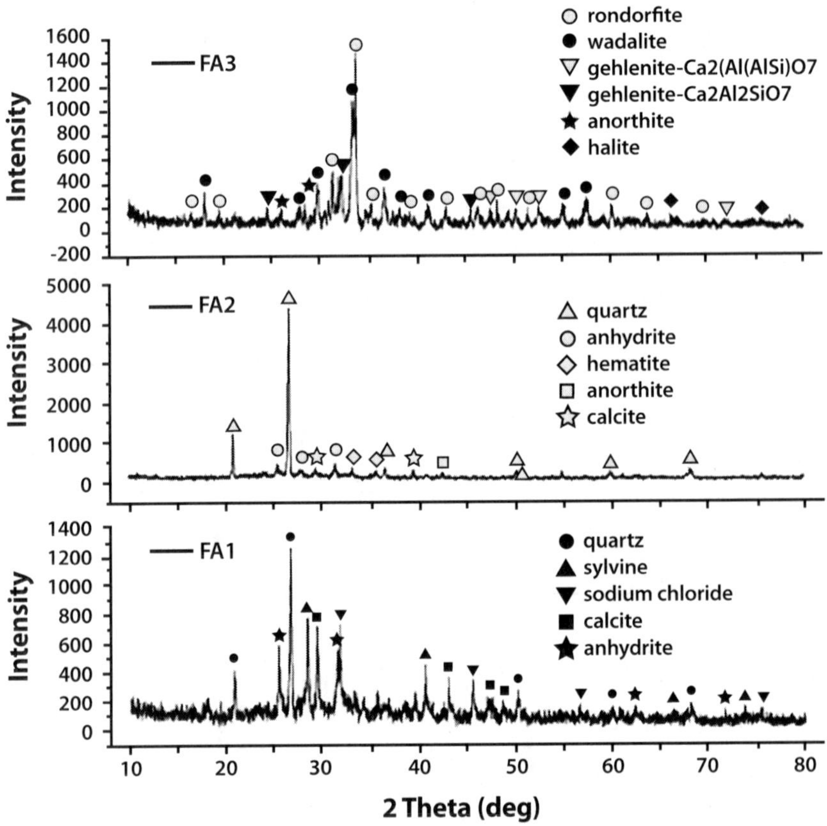

FIGURE 3: XRD results of raw fly ashes.

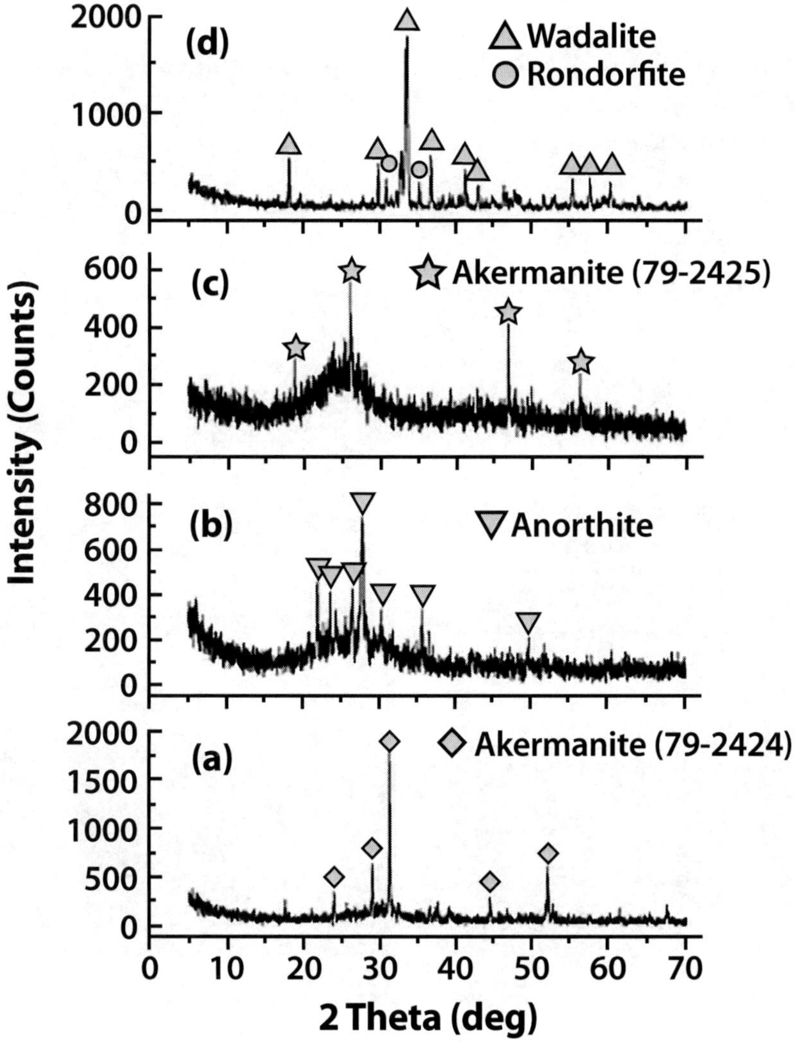

FIGURE 4: XRD results of molten slags. (a)S1; (b) S2; (c) SlagFA3 (FA3+10 wt% SiO_2); (d) S3.

FIGURE 5: SEM of molten slags obtained from various fly ashes (a) S1; (b) S2; (c) SlagFA3 (FA3+10 wt% SiO_2).

The liquid-to-solid ratio was 20:1 and agitation time was 18 hr with rotary tumbler at (30 ± 2) r/min. After extraction, the leachates were examined by AAS (Table 2). The results show that the heavy metals' leaching characteristics of FA1 were very similar to FA3, but as for FA2, the leachability was quite different. For example, the concentration of Zn in FA2 was near to 5 mg/L far exceeding that of FA1 and FA3, and the leaching content of 0.3033 mg/L of Cd was beyond threshold value of Cd according to the Environmental Protection Administration of China.

After melting treatment, TCLP was also adapted to examine the metals' leaching abilities in the produced slags, named S1, S2 and S3, respectively. And the concentrations are listed in Table 2. It is indicated that the slags show well effect on retaining heavy metals. Particularly, we can find that resistance to dissolution was most effective for Cr, Cd and Pb which were highly toxic. Compared with FA2, the extracted amounts of S2 swiftly decreased, especially for Cd, Pb, Cr and Zn. As for S1 and S3, the amounts of several heavy metals were even higher than those in raw ashes, such as Pb, Zn and Cu. The reason was that the high content of Chlorine in FA1 and FA3 impeded the stabilization of heavy metals [25-27] and the high value of basicity of FA1 and FA3 led to poor efficiency of vitrification [28]. It should be noted here, the total elemental mass balances were not considered in this study for some volatile metals such as Cd and Pb may evaporate to the atmosphere during molten stage. Therefore, when using thermal melting technology to treat fly ash, a secondary air pollution control system should be designed to catch volatile metals [29].

Additive for melting: The experiments on improving the melting performance of FA1 and FA3 were conducted by adding SiO_2 (analytic grade) (Western-Union Chem-Industrial Corp., Shanghai) into the raw fly ashes at the proportion of 10 wt. % and 20 wt. %, respectively. Liquid ceramic (LC, composed of SiO_2 and Al_2O_3) (Beijing Dingxin Aihua Science-Technologies Corp.) was also introduced into FA3 by 2.5 wt. %, as another additive for comparison. Then these samples were placed into the plasma melting furnace for vitrification and the vitreous slags, named SlagFA1 and SlagFA3, respectively, were analyzed to evaluate their physical and chemical properties.

The leaching content of heavy metals is showed in Figure 2. As can be seen, for both SlagFA1 and SlagFA3, the leaching amounts decreased significantly when SiO_2 was used as additive to raw fly ashes. For SlagFA1 (Figure 2a), the leaching level of Cu and Zn was noticeably reduced. The amount of Cu dramatically decreased from 1.53 mg/L to 0.09 mg/L (10 wt% SiO_2) and 0.11 mg/L (20 wt% SiO_2). And the amount of Zn remarkably decreased from 5.9 mg/L to 1.26 mg/L (10 wt% SiO_2) and 0.47 mg/L (20 wt% SiO_2), respectively. For SlagFA3 (Figure 2b), the molten slags of blended ashes exhibited much better effect on retaining heavy metals than that of fly ash alone. The immobilizing of Cd, Cr, Pb and Ni was significantly improved, except for Cu and Zn. For Cu, no noteworthy trend of the leaching concentrations could be observed with a variety of SiO_2 values, and for Zn, the leaching value decreased as the portion of SiO_2 increased. While 2.5 wt% of LC was introduced into FA3, the produced molten slag showed extraordinary effect on immobilization of heavy metals. Similar results were reported by other researchers [30,31]. In general, chemical stability is consistent with the progressive formation of a more compact and interconnected glass network structure with the addition of the glass formers. Therefore, addition of SiO_2 and LC strengthened the chemical stability of the glasslike slags.

2.3.2 XRD OF FLY ASH AND SLAG

The X-ray diffraction (XRD) investigations were carried out with a Rigaku Model D/max-rA diffractometer using Cu Kα radiation, operated at 40kV and 100mA in the 2θ range from 5° to 80°. Crystalline phases were identified by comparing intensities and positions of Bragg peaks with those listed in the Joint Committee on Powder Diffraction Standards data files.

Figure 3 shows the XRD results of the raw ash samples. As can be seen, the main phases were quartz (SiO_2), hematite (Fe_2O_3) and calcium salt ($CaSO_4$ and $CaCO_3$) in FA2. This crystal characterization had disadvantage over fastening metals but was conductive to vitrification. As for FA1 and FA3, their crystalline phases were much complex including halite (NaCl), sylvine (KCl) and rondorfite (Cl-bearing) and this kind mineral phase led to poor effect of melting raw materials [32].

The crystal phases of molten slags are exhibited in Figure 4. The XRD pattern of each slag was completely different from that of raw ash. Compared with S1 and S3, both S2 and SlagFA3 show no noticeable crystalline peaks and confirm the amorphous glass structure which contributed to holding heavy metals in the silicate glass framework [33].

2.3.3 SEM OF SLAG

The produced vitrified slags were ground to powder then their microstructure characterizations were investigated using Scanning Electron Microscopy (SEM). The SEM micrographs are shown in Figure 5. These SEM images indicate that these three fly ashes can be transformed into extremely compact and uniform vitreo us slags under appropriate vitrification, thus making them more inert to chemical etching and higher mechanical strength for reclamation.

2.3.4 VOLUME REDUCTION

Archimedes method was utilized to measure the density of raw ashes and molten slags. After thermal melting treatment, the densities of samples significantly increased from 0.62-0.84 g/cm^3 (fly ashes) to 1.89-3.11 g/cm^3 (slags). Thus the reduction of volume reached 60-73%.

2.4 CONCLUSIONS

A lab-scale DC thermal plasma melting system was used to dispose the hazardous fly ashes. After the melting process, the crystalline phases and microstructures of raw fly ashes were changed drastically, and the produced slags exhibited a glass-like monolithic morphology and interconnected compact microstructure, the reduction of bulk volume was in range of 60-73%. Compared with the raw ashes, the molten slags manifest well leaching-resistance of heavy metals, especially for Cd, Pb and Cr. The high chlorine content and low basicity in raw ash hampered solidification of heavy metals. Nevertheless, additive of SiO_2 and LC conduced to the formation of silicate glassy for immobilization of heavy metals and enhanced vitrification treatment significantly.

In conclusion, the thermal plasma torch is an alternative and promising technology for vitrification of hazardous fly ash.

REFERENCES

1. Wiles CC (1996) Municipal solid waste combustion ash: state of the knowledge. Journal of Hazardous Materials 47: 325-344.
2. Sabbas T, Polettini A, Pomi R, Astrup T, Hjelmar O, et al. (2003) Management of municipal solid waste incineration residues. Waste Manag 23: 61-88.
3. Disposal Committee of Urban Domestic Refuse of CAEPI, (2012) China development report on disposal industries of urban domestic refuse in 2012. China Environmental Protection Industry 3: 20-26.
4. Shaaban AF (2007) Process engineering design of pathological waste incinerator with an integrated combustion gases treatment unit. J Hazard Mater 145: 195-202.
5. Alvim-Ferraz MC, Afonso SA (2005) Incineration of healthcare wastes: management of atmospheric emissions through waste segregation. Waste Manag 25: 638-648.
6. Sukandar S, Yasuda K, Tanaka M, Aoyama I (2006) Metals leachability from medical waste incinerator fly ash: A case study on particle size comparison. Environ Pollut 144: 726-735.
7. Zhao L, Zhang FS, Wang K, Zhu J (2009) Chemical properties of heavy metals in typical hospital waste incinerator ashes in China. Waste Manag 29: 1114-1121.
8. Cobo M, Gálvez A, Conesa JA, Montes de Correa C (2009) Characterization of fly ash from a hazardous waste incinerator in Medellin, Colombia. J Hazard Mater 168: 1223-1232.
9. Yan JH, Peng Z, Lu SY, Li XD, Ni MJ, et al. (2007) Degradation of PCDD/Fs by mechanochemical treatment of fly ash from medical waste incineration. J Hazard Mater 147: 652-657.
10. Diaz LF, Savage GM, Eggerth LL (2005) Alternatives for the treatment and disposal of healthcare wastes in developing countries. Waste Manag 25: 626-637.
11. Pan X, Yan J, Xie Z (2013) Detoxifying PCDD/Fs and heavy metals in fly ash from medical waste incinerators with a DC double are plasma torch. J Environ Sci (China) 25: 1362-1367.
12. Zhao P, Ni G, Jiang Y, Chen L, Chen M, et al. (2010) Destruction of inorganic municipal solid waste incinerator fly ash in a DC arc plasma furnace. J Hazard Mater 181: 580-585.

13. Wang Q, Yan JH, Tu X, Chi Y, Li XD, et al. (2009) Thermal treatment of municipal solid waste incinerator fly ash using DC double arc argon plasma. Fuel 88: 955-958.
14. Lin YM, Zhou SQ, Shih SI, Lin SL, Wang LC, et al. (2011) Fate of polychlorinated dibenzo-p-dioxins and dibenzofurans during the thermal treatment of electric arc furnace fly ash. Aerosol and Air Quality Research 11: 584-595.
15. Katou K, Asou T, Kurauchi Y, Sameshima R (2001) Melting municipal solid waste incineration residue by plasma melting furnace with a graphite electrode. Thin Solid Films 386: 183-188.
16. Liu HQ, Wei GX, Liang Y, Dong FY (2011) Glass-ceramics made from arc-melting slag of waste incineration fly ash. Journal of Central South University of Technology 18: 1945-1952.
17. Cheng TW, Tu CC, Ko MS, Ueng TH (2011) Production of glass-ceramics from incinerator ash using lab-scale and pilot-scale thermal plasma systems. Ceramics International 37: 2437-2444.
18. Károly Z, Mohai I, Tóth M, Wéber F, SzépvÃ¶lgyi J (2007) Production of glass-ceramics from fly ash using arc plasma. Journal of the European Ceramic Society 12: 1721-1725.
19. Bonizzoni G, Vassallo E (2002) Plasma physics and technology: industrial applications. Vacuum 64: 327-336.
20. Gomez E, Rani DA, Cheeseman CR, Deegan D, Wise M, et al. (2009) Thermal plasma technology for the treatment of wastes: a critical review. J Hazard Mater 161: 614-626.
21. Pan XC, Yan JH, Ma ZY, Tu X, Cen KF (2008) Research on characteristics of a DC double anode arcs plasma torch. Journal of Power Engineering 28: 132-136.
22. Tu X, Chéron BG, Yan JH, Cen KF (2007) Dynamic behaviour of dc double anode plasma torch at atmospheric pressure. Journal of Physics D: Applied Physics 40: 3972-3979.
23. Tu X, Yan JH, Chéron BG, Cen KF (2008) Fluctuations of DC atmospheric double arc argon plasma jet. Vacuum 82: 468-475.
24. Tu X, Yu L, Yan J, Cen K, Chéron B (2008) Heat flux characteristics in an atmospheric double arc argon plasma jet. Applied Physics Letters 93: 151501.
25. Okada T, Tomikawa H (2012) Leaching characteristics of lead from melting furnace fly ash generated by melting of incineration fly ash. J Environ Manage 110: 207-214.
26. Rio S, Verwilghen C, Ramaroson J, Nzihou A, Sharrock P (2007) Heavy metal vaporization and abatement during thermal treatment of modified wastes. J Hazard Mater 148: 521-528.
27. Jung CH, Matsuto T, Tanaka N (2005) Behavior of metals in ash melting and gasification-melting of municipal solid waste (MSW). Waste Manag 25: 301-310.
28. Takaoka M, Takeda N, Miura S (1997) The behavior of heavy metals and phosphorus in an ash melting process. Water Science and Technology 36: 275-282.
29. Cheng TW (2004) Effect of additional materials on the properties of glass-ceramic produced from incinerator fly ashes. Chemosphere 56: 127-131.
30. Jiang YH, Xi BD, Li XJ, Wang Q, Zhang XX (2005) Effect s of SiO2 on melting and solidification characteristics of fly ash from refuse incinerator. Research of Environmental Sciences 18: 71-73.

31. Li RD, Nie YF, Li AM, Wang L, Chi Y, et al. (2004) [Influence of liquid ceramic additive on binding of heavy metal during the vitrification of fly ash from municipal solid waste incinerator]. Huan Jing Ke Xue 25: 168-171.
32. Tian S, Li J, Liu F, Guan J, Dong L, et al. (2012) Behavior of heavy metals in the vitrification of MSWI fly ash with a pilot-scale diesel oil furnace. Procedia Environmental Sciences 16: 214-221.
33. Cheng TW, Huang MZ, Tzeng CC, Cheng KB, Ueng TH (2007) Production of coloured glass-ceramics from incinerator ash using thermal plasma technology. Chemosphere 68: 1937-1945.

PART II

GASIFICATION

CHAPTER 3

Biomass and RDF Gasification Utilizing Ballistic Heating TGA Analysis

M. J. CASTALDI

3.1 INTRODUCTION

There is a growing recognition that conversion or valorization of wastes (municipal solid wastes (MSW) and agricultural wastes) is an environmentally responsible way of treating the increased volumes without occupying a large portion of land. Many companies are turning to gasification of MSW, biomass and mixtures of these to produce fuels and chemicals.

Zanzi et al. studied biomass decomposition in a free fall reactor operating at different temperatures and residence times (Zanzi et. al. 1996&2002). They measured gaseous emissions coming from pyrolysis of biomass samples. The primary products measured were H_2 between 13–17 %, CH_4 between 12–18 % and CO between 46-53%. In addition, they did detect C_2H_6, C_2H_4 and C_2H_2 but did not resolve them, and thus reported total concentrations that ranged between 4.8–6.3% for C_2H_2 and C_2H_4 combined and 0.2-0.6% for C_2H_6. While their conditions are different from

Castaldi MJ. "Biomass and RDF Gasification Utilizing Ballistic Heating TGA Analysis" Proceedings Venice 2012, Fourth International Symposium on Energy from Biomass and Waste, *Venice, Italy.*

those in the present study, C_2H_6 consistently had the lowest concentration in both studies.

Di Blasi et al. (1999) developed a thin packed bed of biomass and agriculture waste particles to investigate the degradation in relation to heat transfer controlled reactions during gasification. Measurements within the particle bed clearly show a time dependence of approximately 200 seconds to heat the center of the packed bed to the external surface temperature of 900K. Gas measurements were taken as a function of time and show CO_2 and CO evolving earliest with peak production near 100 seconds with 2.75 and 2.0 volume%, respectively. This was followed by the evolution of H_2, CH_4, C_2H_4 and C_2H_6 that had peak production near 250 seconds with 0.005, 0.02, 0.06 and 0.1 volume%, respectively, and a rapid decrease in concentration at about 270 seconds. They did not report any C_2H_2 or higher order gaseous hydrocarbons.

Nunn et al. investigated rapid pyrolysis of sweet gum hardwood and milled wood lignin in an inert reaction atmosphere comprised of helium and argon. The system consisted of an electrically heated screen that achieved heating rates of 1000 K sec^{-1} and enabled measurement of weight loss and gaseous products except hydrogen. The detection of gases occurred from 300oC and higher with production of most gases leveling off near 700oC. In addition to CO_2, CO, methane, and C_2s, Nunn et al. (1985) detected formaldehyde, acetaldehyde and methanol at peak concentrations of 1.5%, 0.8% and 2.0%, respectively.

Dufour et al. (2009) presented work on biomass pyrolysis employing a flow through reactor setup utilizing a "complete analytical system". Their results showed very good mass balances between 92% and 102% for the tested samples. The analyses of the permanent gases included H_2, CO, CO_2, CH_4, C_2H_4 and C_2H_6 as well as total gas flow produced during the reactions. Their system did not have measurements of the biomass samples during reaction, yet did capture nearly all products that evolved, except for water that they indicated was difficult to quantify, and showed chemical species produced as a function of reactor wall temperature.

Here is presented the results of experiments undertaken utilizing a ballistic heating TGA apparatus with simultaneous gaseous product analysis resolved as a function of temperature. The product gases were measured for different reaction atmospheres that simulate different gasification environments, from inert to highly oxidizing and steam atmospheres. The concentration profiles of C_2 intermediate chemical species in relation to one another and to methane and hydrogen provide some insight into the possible reaction sequences that occur on and near the reacting sample surface.

3.2 FUNDAMENTALS OF GASIFICATION

Gasification is a process for converting solid carbonaceous materials to a combustible gas (e.g., H_2, CO, CH_4, CO_2 mixture). The overall objective is to convert these gases into fuels that can be well integrated into current energy technologies. In general, gasification involves the reaction of a solid fuel with a co-reactant at temperatures ranging from 550-1000°C. Co-reactants are introduced in sub-stoichiometric quantities in order to partially oxidize the fuel to CO and H_2 rather than completely oxidizing to CO_2 and H_2O. If an inert gas, such as N_2, is used in place of a co-reactant the process is called pyrolysis. A conceptual schematic of the reaction path is shown in Figure 1. Here, the relative enthalpy change is shown for different gasification environments. Gasification with air or O_2 is an exothermic reaction, so heat is released during the reaction and the products have a lower enthalpy than the reactants. Further combustion of the syngas will form combustion products (CO_2 and H_2O) which have a lower enthalpy than gasification products. The two steps (waste to syngas and syngas to combustion products) have approximately the same enthalpy change. Reaction in an inert environment such as N_2 will require heat input to initiate the reaction. Hence, the products will have a slightly higher enthalpy than the reactants. In this reaction environment hydrocarbons will form and there will be lower conversion to H_2 and CO. Reaction with steam is highly endothermic and heat must be added to the process. The presence of steam results in high concentrations of H_2 which are formed from the water, so the product stream has a higher enthalpy than the reactants. The reaction of CO_2 with waste is also endothermic and the addition of CO (via conversion of CO_2 to CO) creates a higher enthalpy product stream. CO_2 is a more stable molecule than H_2O (heat of formation is -394 kJ/mol for CO_2 and -229 kJ/mol for steam) so the increase in enthalpy is slightly higher for CO_2 gasification than for steam gasification (in other words, combustion of one mole of CO will result in a greater heat release than combustion of one mole of H_2). The conversion of synthesis gas to chemicals or fuels is an exothermic process, and is therefore accompanied by a decrease in enthalpy.

Therefore, gasification can be described as a 3 step process where the first step is dehydration, followed by pyrolysis of volatiles, and finally gasification of char (Bridgwater 2003). This can be understood by observing the mass loss of a solid material during gasification. Steam is commonly used as a co-reactant in gasification processes since it can increase the overall H_2 yield. CO_2 can also be used as a co-reactant during gasification and has been shown to reduce energy requirements to the gasifier as well as increase overall conversion of the solid fuel (Castaldi & Dooher, 2007 & Butterman et. al. 2009).

3.3 EXPERIMENTAL STUDY

3.3.1 APPARATUS

Three samples were provided by Veolia Environment Research Center located in Limay, France. Two samples were a type of refuse derived fuel (RDF) labeled sample A and C and one was a woody biomass sample labeled Clean Wood (CW). The empirical formula and ash percentage of the samples, determined from ultimate analyses prior to testing, are provided in Table 1. Also included in Table 1 is the amount of oxygen that is required for complete combustion of the sample. As will be discussed in the Results and Discussion section, complete combustion does not occur, but this value provides context for the amount of oxygen being added during the testing.

The sample "A" is made from a preparation of raw municipal solid waste via sorting, shredding, typical separation steps to extract the organic fraction of MSW. Sample C is made from a preparation of raw industrial solid waste with an equal distribution of paper, plastics, wood, and cardboard. These samples have been removed from an existing industrial scale facility load preparation line using a pre-determined sampling procedure, and then dried and micronized by pilot scale equipment. Sample CW is made from untreated wood chips to compare to the waste samples and as an independent data set for biomass.

TABLE 1: Empirical Formula of Tested Samples

	Sample A	Sample C	Clean Wood CW
Empirical Formula	$CH_{1.8}1O_{0.62}N_{0.02}S_{0.001}$	$CH_{1.56}O_{0.62}N_{0.01}S_{0.002}$	$CH_{1.42}O_{0.61}N_{0.003}S_{0.0002}$
Ash % (Dry Basis)	17.70	14.67	0.60
O_2 for complete combustion (moles)	0.759	0.775	0.830

The samples were processed in the gasification test facility of the Combustion and Catalsyis Labortory. The equipment consists of an Instrument Specialists Temperature Programmer Interface/Thermal Analyzer that can regulate the temperature and heating rate of the quartz furnace in the Dupont 951 Thermogravimetric Analyzer. The carrier flow consists of N_2 (UHP grade, TechAir, New York), O_2 (HP grade, Tech Air, New York) Ar (UHP grade, TechAir, New York) and Air (Bone dry grade, TechAir, New York) whose flow rates are regulated by means of Gilmont rotameters (model # GF 1060) or Aalborg mass flow controllers (Model # GFC17). A syringe pump (model kd-Scientific 780-100) feeds distilled water into a stainless steel steam generator that produces slightly superheated steam (~110-120oC) whose temperature is monitored by an in-line thermocouple with digital readout (model Omega E-type) prior to entering the furnace.

The products of gasification exit the furnace and are sampled using a rapid syringe capture technique that condenses moisture from the gas evolution products for subsequent GC analysis. The typical size of the solid samples ranged from 20–25 mg. The sample weights were measured in real time by the TPI/TA software with verification by means of a Mettler scale accurate to +/- 0.1mg. Gas analysis was conducted with an Agilent 3000 Micro GC with a total run time of about 3 minutes for each syringe sample analysis. The instrument is compatible with mixtures in a gaseous phase at STP, typically for compounds with boiling points below 250°C.

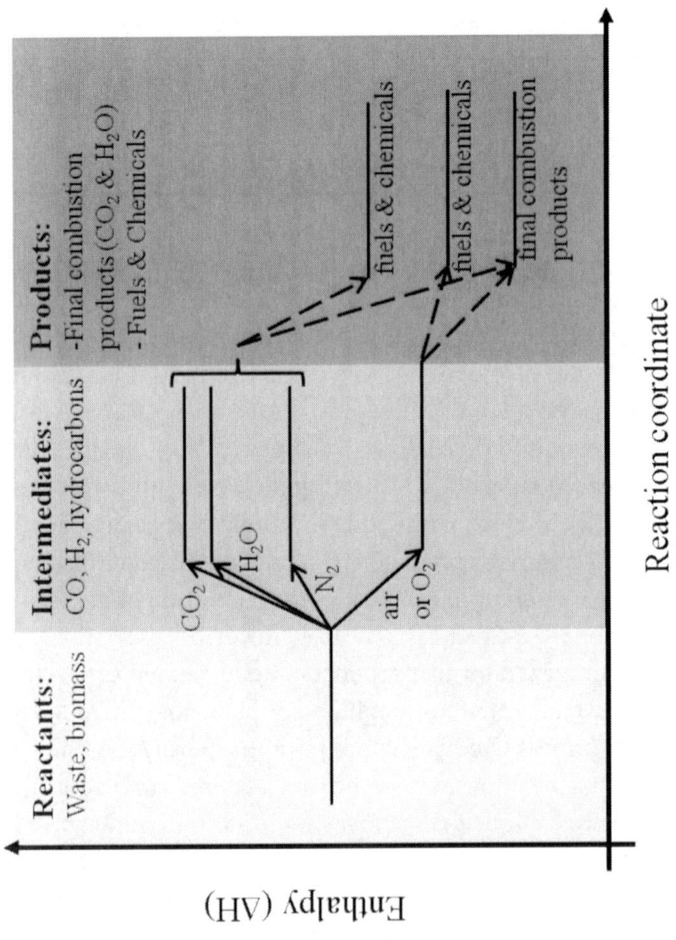

FIGURE 1: Conceptual pathway for conversion of solid fuel to different products. Enthalpy change is shown for different co-reactants.

The minimum detection limit has been established to be about 1–5 ppmV for permanent gases with a linear dynamic range of 10 ppm to 100%. The Micro GC has the capability of measuring a wide range of species including C_2-C_5 hydrocarbons and oxygenated hydrocarbon species. Those species that were measured in the current study were O_2, N_2, H_2, CO, CO_2, CH_4, C_2H_6, C_3H_8, C_4H_{10}, C_2H_4, C_3H_6, and C_2H_2. The columns installed in the currently used Micro GC are not capable of measuring H_2S or H_2O levels.

3.4 RESULTS AND DISCUSSION

Figure 2 shows a comparison of the sample A mass decompositions for two different stoichiometric amounts of air. Even though the rates of decay are rapid, there is a slight difference as would be expected for the given amount of air. For example, comparing the weight fractions (W_{300}/W_o) at 300°C, sample A with an air equivalence ratio of 0.6 has a W_{300}/W_o = 49.7% whereas with an equivalence ratio of 0.2, W_{300}/W_o = 52.9%. Therefore, the higher stoichiometry test resulted in a more rapid pyrolytic decomposition rate, as expected. The remaining mass decomposition curves will not be shown as they are similar to sample A results. While the rates of decomposition are rapid, due to the rapid heating rate, it is possible to discern faster or slower reaction progressions with different amounts of reacting atmospheres.

Figure 3 shows the major product gases from the gasification of sample A as a function of measured sample temperature. Again as a representative data set, the gaseous chemical species that evolved during the test from sample A in Air with a furnace pre-set of 1000oC are shown completely in Figure 3. The data sets for the samples C and Clean Wood behave similarly. For this test the Air flow rate was set to 2800 Nl (kg dry product)-1 providing an equivalence ratio of 0.6. The behavior of the reactant and product evolution follows expected trends. At lower temperatures there is more partial oxidation and at higher temperatures there is more complete oxidation occurring.

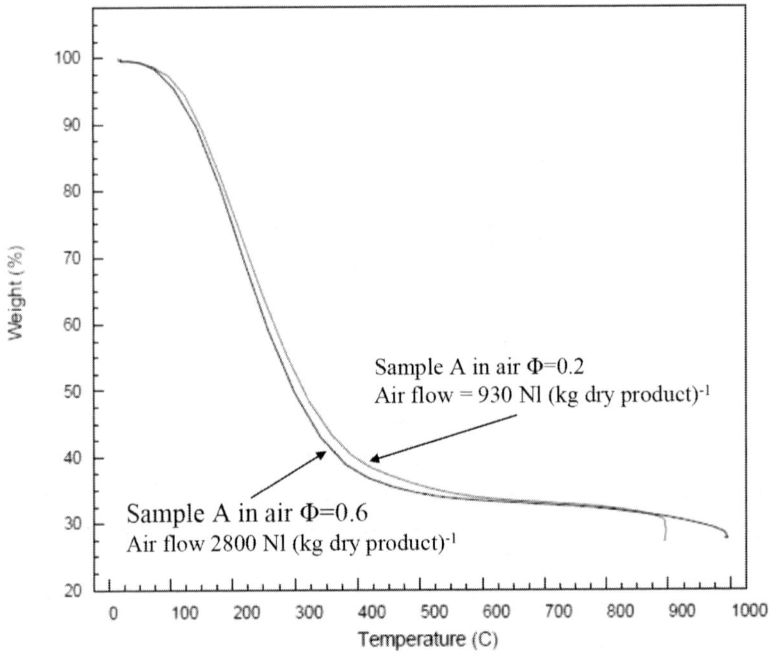

FIGURE 2: Direct comparison of mass decomposition for sample A in Air, ambient to 1000°C.

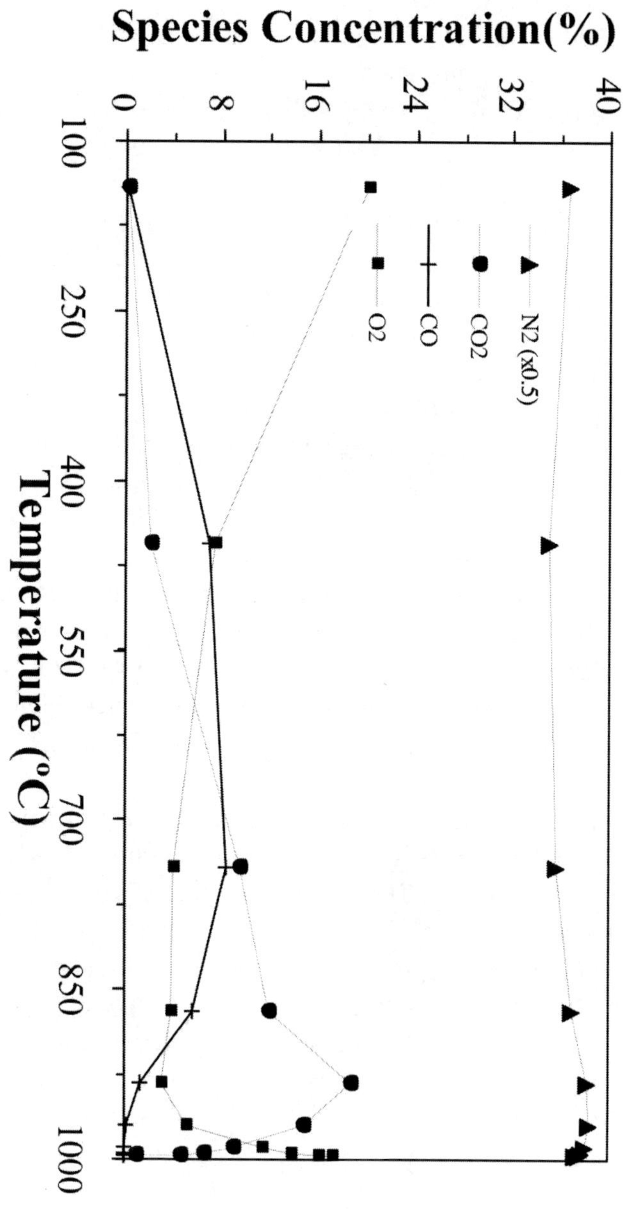

FIGURE 3: Major product species profiles for test sample A in Air (%Volume), 1000oC test (Φ=0.6).

Shown in Figure 3, the N_2 concentration remains nearly constant at 74.6%, which is about 5% lower than the expected 78.5% based on the input flow rates. This difference can be attributed to non-condensable species produced but not analyzed by the GC combined with a gas expansion and product evolution. Notice this first point is taken when the sample is at 100°C which is a result of the testing configuration and timing for the sample introduction into the furnace. Here some oxygen is consumed and products are being evolved. As the sample is further heated, the O_2 decreases to about 5% and the production of CO_2 and CO is observed. Initially, below 400°C, the production of CO is greater than CO_2 indicating a combination of devolatilization processes combined with partial oxidation. The rapid heating rate employed likely produces a significant amount of volatile hydrocarbons in a very short period. This would prevent O_2 from reaching the solid surface because there will be a large amount of hydrocarbons in the vapor phase just above the surface that will react with the oxygen attempting to diffuse to the surface. Since the temperature is near 400°C and the amount of hydrocarbons would be large the CO observed is likely due to primary reactions within the biomass. This has been observed in other studies where the initial stages of biomass gasification and pyrolysis yielded CO and CO_2 (Nunn 1985, Dufour 2009). This is consistent with hydrocarbon evolution observed in Figure 4 where higher hydrocarbons are observed mainly for lower temperatures. However, CO_2 concentration becomes higher than CO at temperatures greater than 750°C. This is due to oxidation reactions of primarily volatile hydrocarbons and CO.

Figure 4 shows the gaseous product evolution of the minor species as the reaction proceeds. Those chemical species identified range from methane to butane.

As observed in Figure 4, all the minor species increase and reach their peak production near 400°C. This is consistent with Figure 3 in that oxygen is becoming consumed and partial oxidation and pyrolytic decomposition is occurring. As the sample temperature increases and the volatiles are driven off or have reacted, the minor species concentrations decline. Here it is likely that any oxygen remaining in the sample and the oxygen from the injected air is sufficiently reactive to completely oxidize all of the remaining hydrocarbon species present.

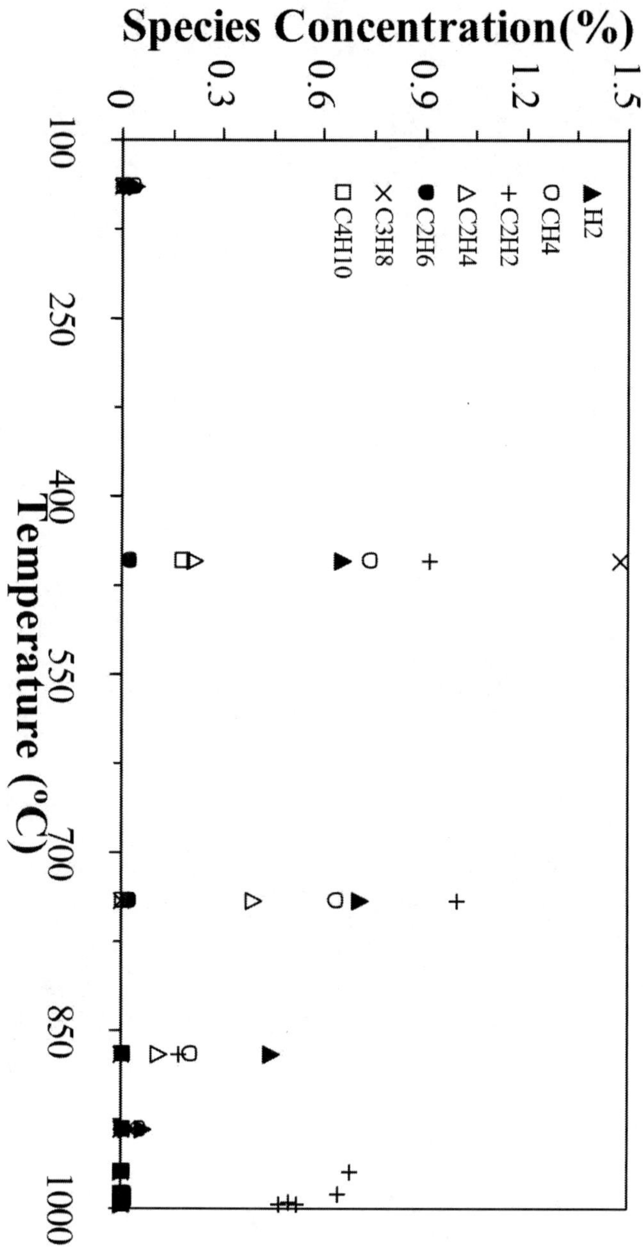

FIGURE 4: Minor product species evolution for test sample A in Air (%Volume), 1000°C test (Φ=0.6).

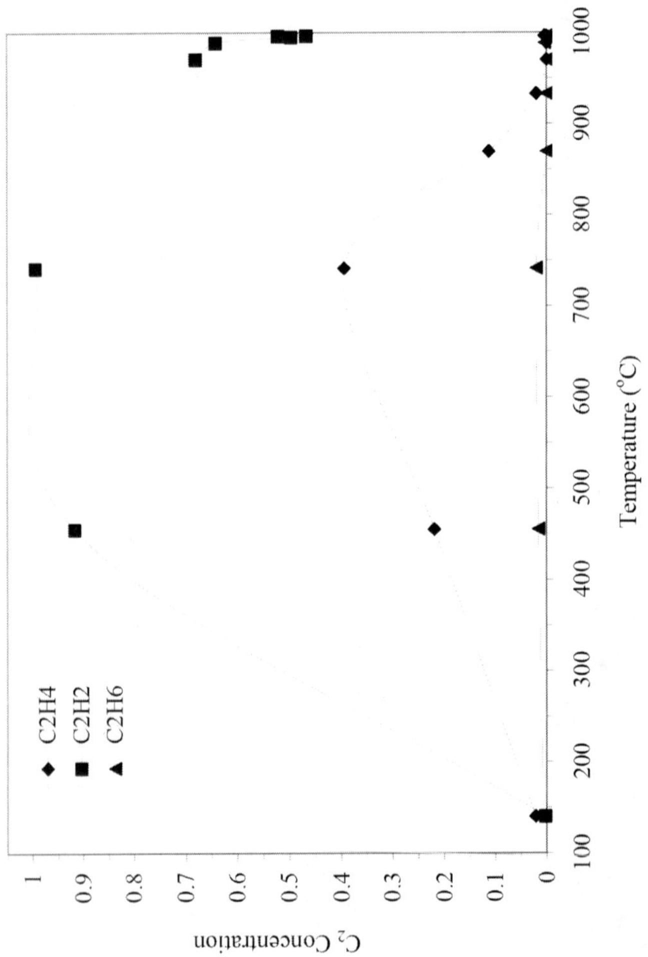

FIGURE 5: Concentration profile of C_2 chemical species for sample A in Air (%Volume), 1000oC test (Φ=0.6).

Figure 5 shows the observed trends for the production of C_2 species as a function of temperature. This data is the same as shown in Figure 4. As can be seen, the acetylene concentration is slightly lower at temperatures below 200°C but then rapidly increases and persists until the sample is exhausted. This is consistent with the production of high hydrogen to carbon volatiles that are dehydrogenated homogeneously as they are oxidized and with lower hydrogen to carbon compounds evolved as temperature increases and more comes directly from the sample surface. Typically a temperature (or time) dependent relationship is observed for the transition from ethane to acetylene. For example, as ethane is consumed a commensurate rise in ethylene concentration would be observed. This would be followed by a decrease in ethylene concentration, at higher temperatures, and a commensurate increase in acetylene. This exact relationship cannot be observed here likely due to the rapid heating rate and the presence of a solid sample, yet the general trend is retained. The acetylene can be observed to have a slightly lower concentration below 200°C yet start to decrease in concentration at a temperature close to 900°C where the ethylene is already rapidly decreasing and the ethane is completely consumed. The region between 300°C and 750°C produces increased amounts of all three compounds likely due to the abrupt evolution of material from the sample surface, thus overwhelming the C_2 reaction mechanism discussed above.

Once formed, acetylene is difficult to oxidize thus it penetrates into the TGA exhaust.

At higher temperatures (>500°C) the mass decomposition profiles decay at a similar rate unlike what occurs at lower pyrolysis temperatures (<350°C) where the Clean Wood undergoes an earlier steep decline in mass. At these higher temperatures, area under the concentration curves can be correlated with relative gas evolution quantities. This indicates that feedstock A having a higher H/C ratio is correlated with the alkane product CH_4 (H/C = 4) having a higher H/C ratio.

Figures 6 and 7 show the comparisons for methane and ethylene as a function of temperature and test sample in a steam/Ar reaction environment. The order for H/C ratio is A > C > CW. It appears that the methane and ethylene are correlated to H/C ratio in that higher ratios yield high concentrations of these hydrocarbons.

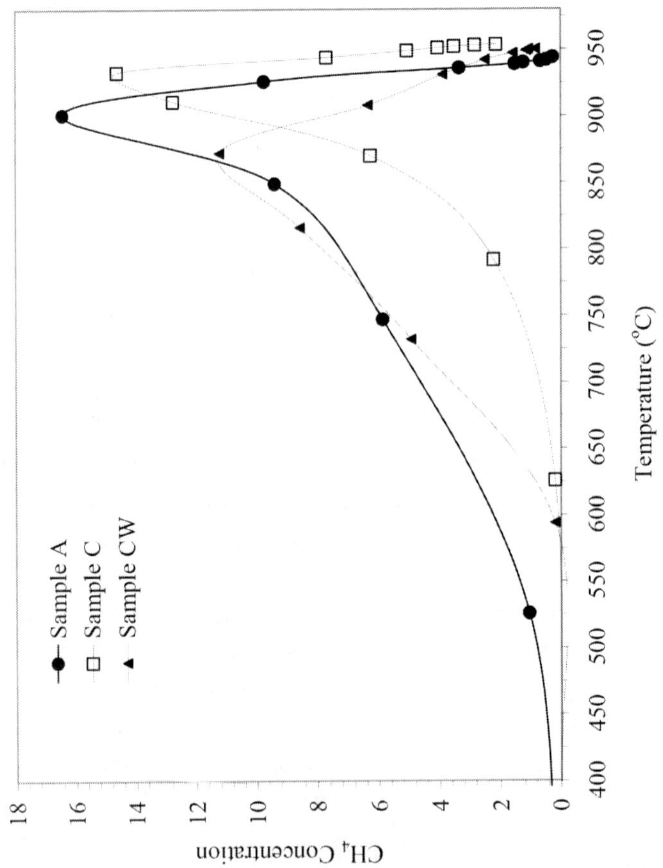

FIGURE 6: Methane concentration profiles (%Volume) as a function of reaction temperature comparing sample A (H/C = 1.81), C (H/C = 1.56) & CW (H/C = 1.42) in a steam/Ar reaction atmosphere.

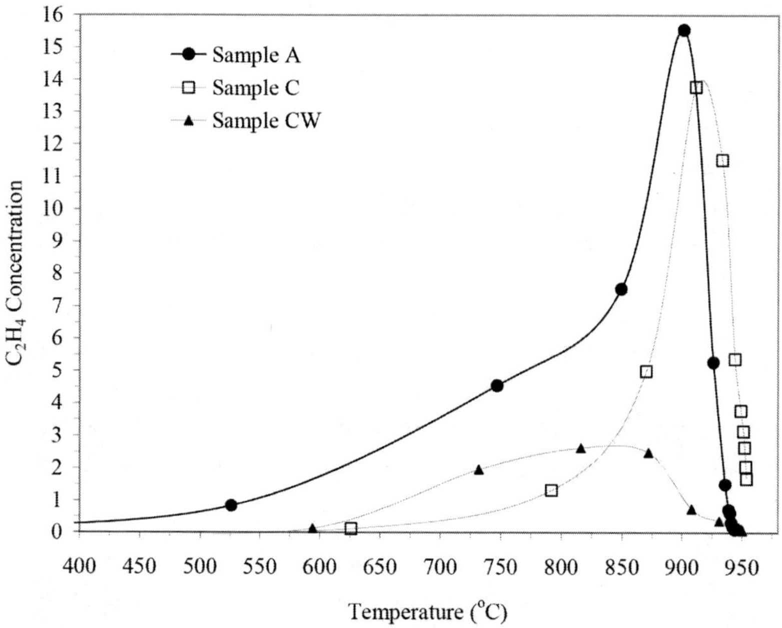

FIGURE 7: Ethylene concentration profiles (%Volume) as a function of reaction temperature comparing sample A (H/C = 1.81), C(H/C = 1.56) & CW (H/C = 1.42) in a steam/Ar reaction atmosphere.

To determine the apparent kinetic parameters from the mass decay curves a percent fit (% error minimization) determination of kinetic parameters through a minimization of the squared deviations between the calculated and measured weight loss fraction derivatives was used. The best fit was determined to be a global first order decay, i.e. $n = 1$ through a comparison of error terms resulting from testing reaction orders $n = 1$, 2 and 3. Values of the pre-exponential factor A and activation energy Eact for $n = 1$ are presented in Table 2. Kinetics at 1000°C could not be extracted since, at the steeper heating rates, noise in the higher order derivatives was extremely high and rendered the calculations unreliable.

A more reactive sample would begin decomposition earlier and possess a lower global activation energy. Nevertheless, variability in chemical structure may result in a sample that is more thermally resistant requiring a wider temperature range or longer residence time for thermal processing. Here, variability refers to differences in molecular structure (such as lignin having different methoxy to phenylpropanoid ratios or extent of crystallinity of the cellulose fraction) rather than the macroscopic structure of the biomass that would involve the xylem/phloem/fibrous/porous nature of the sample.

The hemicellulose pyrolysis peak occurs at about 270°C, cellulose at about 360oC and the polypropylene/polyethylene plastics at about 460–480°C:

$$Eact_{hemi}(\sim 40 \text{ kJ/mole}) < Eact_{lig}(\sim 60) < Eact_{cell}(\sim 210) < Eact_{pp/pe}(\sim 240)$$

The slope of the pyrolysis curve is not necessarily correlated with the activation energy—while the cellulose component of biomass has a much steeper curve and more narrow decomposition interval, its activation energy is much greater than that of lignin whose decomposition starts earlier and is spread over a much wider temperature range. These are consistent with findings from other groups that determined kinetics during fast reaction processes (Grammelis et. al. 2009, Varhegyi et. al. 1994, Yang et. al. 2007). Table 2 shows the averaged values of the parameters estimated from the mass decay curves. There was significant scatter in the pre-exponential factors, but the apparent activation energies were more consistent.

TABLE 2: Average apparent kinetic parameters from mass decay curves

Sample	Pre-Exponential Factor (sec^{-1} $K^{1/2}$)	Apparent Activation Energy (kJ mol^{-1})
CW	$8.00x10^{27}$	22
C	$2.02x10^{29}$	71
A	$3.71x10^{23}$	185

While the composite mass decomposition curve showed the decay of Sample A commencing first followed by C and then CW in the pyrolytic range (lower temperatures), the slope of the decomposition curve was greatest for CW and smallest for A resulting in near complete processing of CW occurring before C and lastly followed by A. This would be expected if Sample A had the highest and CW had the lowest activation energy, which is consistent with the calculation results.

3.5 CONCLUSIONS

A series of Refuse Derived Fuels and Clean Wood were tested in a TGA using heating rates near 700°C min^{-1}. Different reaction atmospheres (argon, Air, 5% and 10% O_2, and steam) enabled important comparisons to be made between the samples. Permanent gases were quantified as a function of temperature and revealed expected behavior with regard to reaction atmosphere.

Concentration profiles of higher order hydrocarbons were presented which provided insight into the likely reaction sequences occurring during the gasification of the samples. Volatile hydrocarbon products in an oxygen reactive environment are dependent on temperature and oxygen concentration rather than the H/C ratio of the sample. Across all feedstocks with increasing O_2 concentration (0-20%), higher levels of CO_2 and C_2H_2

and lower levels of H_2, CH_4 and C_2H_6 were observed. The increased CO_2 levels are explained by greater combustion reaction activity. Higher acetylene levels were produced at the expense of methane and ethane with the production of water instead of H_2. Based on the acetylene concentration profiles it appears the oxygen reacts homogeneously with other hydrocarbons at low temperature and the acetylene does not seem to be released from the samples directly. In a steam environment the hydrogen content in the volatile hydrocarbons is related to the H/C ratio of the sample, and as the H/C ratio increases, methane and ethylene production increase.

Kinetic data were extracted for the different samples and show that Clean Wood was slightly more reactive than RDF from Industrial Solid Waste and more reactive than RDF from Municipal Solid Waste (MSW) and that oxygen bound in the sample matrix is significant in terms of reaction sequence. This study also demonstrates good potential for H_2 production through gasification of RDF (from Industrial and Municipal Solid Wastes) in comparison to Clean Wood gasification.

REFERENCES

1. Bridgwater A.V., Chemical Engineering Journal, 91 (2003) 87-102.
2. Butterman H.C., Castaldi M.J., Environmental Science & Technology, 43 (2009) 9030-9037
3. Castaldi M.J., J.P. Dooher, Int. J. Hydrogen Energy, 32 (2007) 4170-4179.
4. Di Blasi, C.; Signorelli, G.; Di Russo, C.; Rea, G. Product Distribution from Pyrolysis of Wood and Agricultural Residues. Ind. Eng. Chem. Res. 1999, 38, 2216.
5. Dufour, A.; Girods, P.; Masson, E.; Rogaume, Y.; Zoulalian, A. Synthesis gas production by biomass pyrolysis: Effect of reactor temperature on product distribution. Int. J. Hydrogen Energy. 2009, 34, 1726
6. Grammelis, P.; Basinas, P.; Malliopoulou, A.; Sakellaropoulos, G. Pyrolysis kinetics and combustion characteristics of waste recovered fuels. Fuel. 2009, 88, 195.
7. Nunn, T.R.; Howard, J.B.; Longwell, J.P.; Peters, W.A. Product compositions and kinetics in the rapid pyrolysis of milled wood lignin. Ind. Eng. Chem. Process Des. Dev. 1985, 24, 844.
8. Nunn, T.R.; Howard, J.B.; Longwell, J.P.; Peters, W.A. Product compositions and kinetics in the rapid pyrolysis of sweet gum hardwood. Ind. Eng. Chem. Process Des. Dev. 1985, 24, 836
9. Varhegyi, G.; Szabo, P.; Antal, Jr. M.J. Kinetics of the thermal decomposition of cellulose under the experimental conditions of thermal analysis. Theoretical extrapolations to high heating rates. Biomass Bioenergy. 1994, 7, 69.

10. Yang, H.; Yan, R.; Chen, H.; Lee, D.H.; Zheng, C. Characteristics of hemicellulose, cellulose and lignin pyrolysis. Fuel. 2007, 86, 1781.
11. Zanzi, R.; Sjostrom, K.; Bjornbom, E. Rapid high-temperature pyrolysis of biomass in a free-fall reactor. Fuel. 1996, 75, 545.
12. Zanzi, R.; Sjostrom, K.; Bjornbom, E. Rapid pyrolysis of agricultural residues at high temperature. Biomass Bioenergy. 2002, 23, 357.

CHAPTER 4

Cogeneration of Renewable Energy from Biomass (Utilization of Municipal Solid Waste as Electricity Production: Gasification Method)

MISGINA TILAHUN, OMPRAKASH SAHU, MANOHAR KOTHA, AND HEMLATA SAHU

4.1 INTRODUCTION

Biomass is one of the renewable and potentially sustainable energy sources and has many possible applications varying from heat generation to the production of advanced secondary energy carriers. It has almost zero or very low net CO_2 emission since carbon and energy are fixed during the biomass growth [1]. There are different types of technologies for converting biomass to electricity or to a secondary fuel such as thermal conversion, chemical conversion and bio-chemical conversion [2]. However, thermo-chemical methods such as gasification have a great potential in producing a syngas mainly composed of H_2 and CO with traces of different gases such as CH_4 in different proportions [3]. The produced fuel gas can be flexibly utilized in boilers, engines, gas turbines or fuel cells [4]. Smaller scale gasification systems with internal combustion engines can now be used for thousands of hours to give reasonably high electrical efficiencies and limited emissions [5]. However, fuel cells have the potential to operate at higher electrical efficiency and with lower emissions compared with traditional power generation techniques. Fuel cells are emerging as a leading alternative technology to the more polluting internal combustion engines in vehicle and stationary distributed energy applications. In addition, the future demand for portable electric power supplies is likely to exceed the capability of current battery technology. Hydrogen-powered fuel cells emit only water and have virtually no pollutant emissions, even nitrogen oxides, because they operate at temperatures that are much lower than internal combustion engines [6]. However, even fuel cells fuelled by hydro- carbon fuels have the potential to provide efficient, clean and quiet energy conversion, which can contribute to a significant reduction in greenhouse gases and local pollution. When heat generated in fuel cells is also utilized in combined heat and power (CHP) systems, an overall efficiency of 85 % in excess can be achieved [7]. Different types of fuel cells suitable for several energy applications at varying scales have been developed, but all share the basic design of two electrodes (anode and cathode) separated by a solid or liquid electrolyte or a membrane. Hydrogen (or a hydrogen-containing fuel) and air are fed into the anode and cathode of the fuel cell, and the electrochemical reactions assisted by catalysts take place at the electrodes [8]. The electrolyte enables transport of ions between the electrodes while the excess electrons flow through an external circuit to provide electrical power. Fuel cells are classified according to the nature of their electrolyte, which also determines their operating temperature, the type of fuel and a range of applications [9]. The electrolyte can be acid, base, salt or a solid ceramic or polymeric membrane that conducts ions.

Gasification is a process that converts organic or fossil based carbonaceous materials into carbon monoxide, hydrogen and carbon dioxide. Hydrogen and fuel cells are often considered as a key technology for future sustainable energy supply. Renewable shares of 36 % (2025) and 69 % (2050) on the total energy demand will lead to hydrogen shares of 11 % in 2025 and 34 % in 2050 [10]. Today, hydrogen is mainly produced from natural gas via steam methane reforming, and although this process can sustain an initial foray into the hydrogen economy, it represents only a modest reduction in vehicle emissions as compared to emissions from current hybrid vehicles [11]. Biomass has been recognized as a major world renewable energy source to supplement declining fossil fuel resources [12, 13]. It will play an important role in the future global energy infrastructure for the generation of power and heat, but also for the production of chemicals and fuels. The dominant biomass conversion technology will be gasification, as the gases from biomass gasification are intermediates in the high-efficient power production or the synthesis from chemicals and fuels. Biomass gasification offers the earliest and most economical route for the production of renewable hydrogen. International Institute for Applied Systems Analysis (IIASA's) Environmentally Compatible Energy Strategies (ECS) project has developed a long-term hydrogen-based scenario (B1-H2) of the global energy system to examine the future perspectives of fuel cells [14].

The scenario illustrates the key role of hydrogen in a long-term transition towards a clean and sustainable energy future. According to this scenario, biomass gasification will become a dominant technology in the future. The main aim of the work is to utilize the sewage sludge for the production of hydrogen and uses it for electricity generation. This work concentrated on the percentage of hydrogen production with variation of temperature, pressure and residence time by gasification method. Effect of partial pressure of produced hydrogen, working temperature and electrolyte concentration on cell performance has also been studied.

4.2 MATERIALS AND METHODS

4.2.1 MATERIAL

The material was arranged from solid waste dumping area of Kombolcha town (Ethiopia).

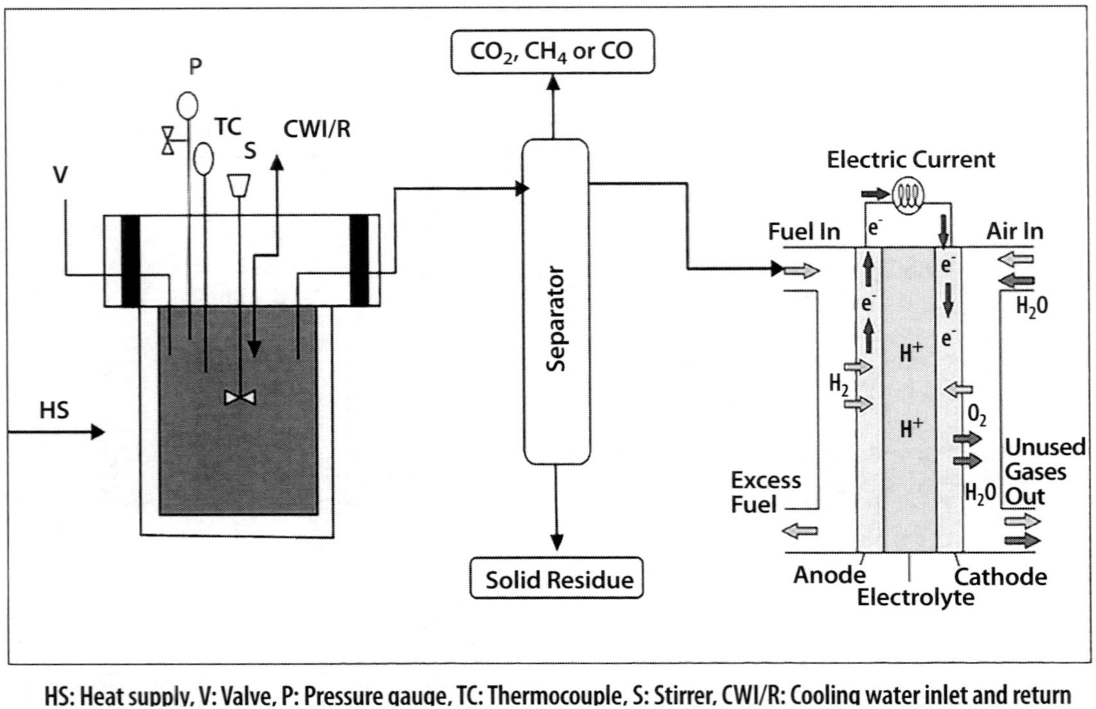

FIGURE 1: Schematic diagram of experimental setup. HS Heat supply, V Valve, P Pressure gauge, TC Thermocouple, S Stirrer, CWI/R Cooling water inlet and return

4.2.2 REACTOR SET UP

The reactor vessel was dual-shell type with an insert made of titanium, widely utilized as corrosion-resistant metal, and a pressure shell. The reactor used in this study was also equipped with auxiliaries such as a stirrer, thermocouples, nozzles, and a pressure gauge. The reaction was initiated by immersing the reactor into molten salt bath (mass ratio of salt was adjusted to K_2NO_3:$NaNO_2$:$NaNO_3$ = 6:5:1). After lapse of predefined time, the reactor was taken out of the bath and subsequently quenched to stop the reaction. HPLC high-pressure pump was used for feeding the distilled water to the reactor to adjust the reaction pressure precisely. The reaction temperature was measured by K-type thermocouple and pressure with digital pressure gauge. The reactor was loaded with deionized water and initial sewage sludge (2 wt % of deionized water) for every experiment (250 rpm and particles size 180 μm). The amount of catalyst was 20 wt % of the organic waste. Then, the air in the reactor was replaced with argon gas. The reactor was sealed and put into the sand bath heated at reaction temperature. It took about 3 min for the reactor to reach the setting reaction temperature around 700 °C. It took about 2 min for final setting of the reaction pressure and reaction time will be considered as zero. As the reaction pressure increased by about 1 MPa than the initial reaction pressure for all experiments, the reaction pressure was assumed to be the initial reaction pressure of the experiment.

4.2.3 GAS ANALYSIS

Produced gas was sampled from one of the sampling loop ports using a gas-tight syringe for gas analysis injection. Liquid and solid residues were collected as mixtures subsequently separated by centrifugation run at 2,500 rpm for 5 min. Moreover, the liquid phase was filtered by 0.45 μm pore size syringe filter (Millex LH, Millipore) and diluted by deionized water prior to the analysis. Separated solid (small amount) residues were dried in an oven kept at 105 °C for at least 6 h and weighed. Gas analysis was carried out with gas chromatograph (GC) GC-2014; SHMADZU equipped with Shin-carbon ST 50/80 column and thermal conductivity detector (TCD) to separate H_2, CH_4, CO, and CO_2. As for the ICP analysis of initial sewage sludge, acid decomposition by nitric acid and sulfuric acid was conducted under 210 °C using an electric hot plate [15]. Guaranteed grade of potassium hydroxide provided by Wako Pure Chemicals Industries, Ltd. was used as a catalyst for gasification.

4.2.4 PREPARATIONS OF ELECTRODE

The anode electrode was prepared by first dispersing the required quantity of catalyst powder in a Nafion® dispersion (SE-5112) for 30 min. An ultrasonic water bath was used to prepare catalyst slurry. The Nafion® dispersions have both hydrophilic and hydrophobic features. Polytetrafluoroethylene (PTFE) is hydrophobic and when employed as a binder, it may prevent hydrophilic fuel from reaching the catalyst site. Therefore, Nafion® has been used to bind the catalyst particles on to the carbon paper. The catalyst slurry was spread on carbon paper in the form of a continuous wet film using a paint-brush technique. It was then dried in an oven for 30 min at 80 °C. Nickel meshes were used as a current-collector because of its non-corrosive nature in an alkaline medium. The catalyzed carbon paper was pressed on to the nickel mesh with application of the Teflon® dispersion. The prepared electrode was pressed at 50 kg cm^{-2} and 120 °C for 5 min to form a composite structure. The area of the working electrode was 25 cm^2. Finally, the composite was heated at 573 K for 4 h to obtain the final form of the anode electrode [16]. Similarly, we have used magnesium oxide for cathode electrode. The Teflon-coated side of the electrode was exposed to the air-side in alkaline fuel cell, and thereby prevented leakage of electrolyte to the air-side and allowed oxygen to permeate.

4.2.5 EXPERIMENTAL SETUP

The experiments were carried out in a 7 cm × 7 cm stainless steel plate in which a special new designed electrolyte carrier plate is fitted with bolts. The cathode (5 cm × 5 cm) and anode (5 cm × 5 cm) are placed in front and back side of electrolyte carrier. A wire connected with the anode and the cathode is used as terminals for measuring current and voltage of the alkaline fuel cell. The space between the anode and the cathode was filled with electrolyte (KOH) with the help of peristaltic pump. The electrolyte was feeded 1 ml min^{-1}, such that one side of the cathode was in contact with the electrolyte and the other side was exposed to air. Oxygen present in the air acts as an oxidant. The hydrogen gas which was generated from the biomass by gasification method is used as anode feed and oxygen is taken from the air. The electrolyte in the beaker was continuously stirred by a magnetic stirrer to maintain a uniform concentration and temperature in the beaker and to reduce any concentration polarization near the electrodes. The voltage and current were measured after a steady state is reached. The complete experimental setup for the production of hydrogen (reactor) and supply for electricity generation (fuel cell) is shown in Fig. 1.

4.3 RESULTS AND DISCUSSION

4.3.1 EFFECT OF TEMPERATURE, PRESSURE AND RESIDENCE TIME

Temperature, pressure, and residence time have been noted to be the most important variables for modifying supercritical reaction conditions [17]. Optimal supercritical conditions can be experimentally derived and aided by models to induce the ideal combination of temperature, pressure, and residence time [18]. System optimization, however, involves maximizing the desired output (energy or organic destruction), while reducing reaction times to minutes or seconds versus the hours required for similar results in subcritical water [19]. The effect of temperature, pressure and residence time is shown in Fig. 2a–c.

4.3.2 EFFECT OF TEMPERATURE

At the chemical equilibrium state, the yields of H_2 and CO_2 increase with the increasing temperature, but the yield of CH_4 decreases sharply. The equilibrium CO yield is very small, and it is about 10^{-3} mol/kg. As temperature increases from 400 to 800 °C, the CO yield firstly increases and then drops down. The maximum CO yield is reached at about 550 °C. Hydrogen yield increases at a low speed at rather higher temperature. When the reaction temperature is above 650 °C, biomass gasification goes to completion and the equilibrium gas product consists of H_2 and CO_2 in a molar ratio equal to (2−y + x/2) (x and y are the elemental molar ratios of H/C and O/C in biomass, respectively). The maximal equilibrium H_2 yield 88.623 % mol/kg of wet biomass was obtained, which is shown in Fig. 2a. From the viewpoint of thermodynamics, higher temperature is essential for hydrogen production. Temperature is considered the most sensitive variable in SCWG processes, with 600 °C serving as an often-cited, optimal target temperature due to associated high conversion [20–22].

4.3.3 EFFECT OF RESIDENCE TIME

At the chemical equilibrium state, the yields of H_2 and CO_2 increase with the increasing residence time, but the yield of CH_4 and CO decreases sharply, which is shown in Fig. 2b. Initially H_2 and CO_2 were 49 and 23 % of the yield which is slow; and after 30 min they gradually increased and at 60 min, they reached up to 88.8 and 50 %. But in the case of CH_4 and CO it decreased from 27 and 16 % to 10^{-3} mol/kg. Longer residence time can improve gasification thoroughness, but there is also an inverse relationship between temperature and reaction completeness, dropping from a few minutes below 600 °C to a few seconds above 600 °C. The optimal temperature threshold for SCWG (i.e., 600 °C) has been shown on the lower side of the conversion range for higher concentration biomass in the absence of a catalyst [23–26]. From the viewpoint of thermodynamics, biomass can be gasified completely in SCW with a product formation of H_2 and CO_2, but adequate reaction time was required to complete the gasification process. Short residence times and high organics destruction efficiencies occur during gasification and oxidative reactions at supercritical operating conditions above 600 °C [27].

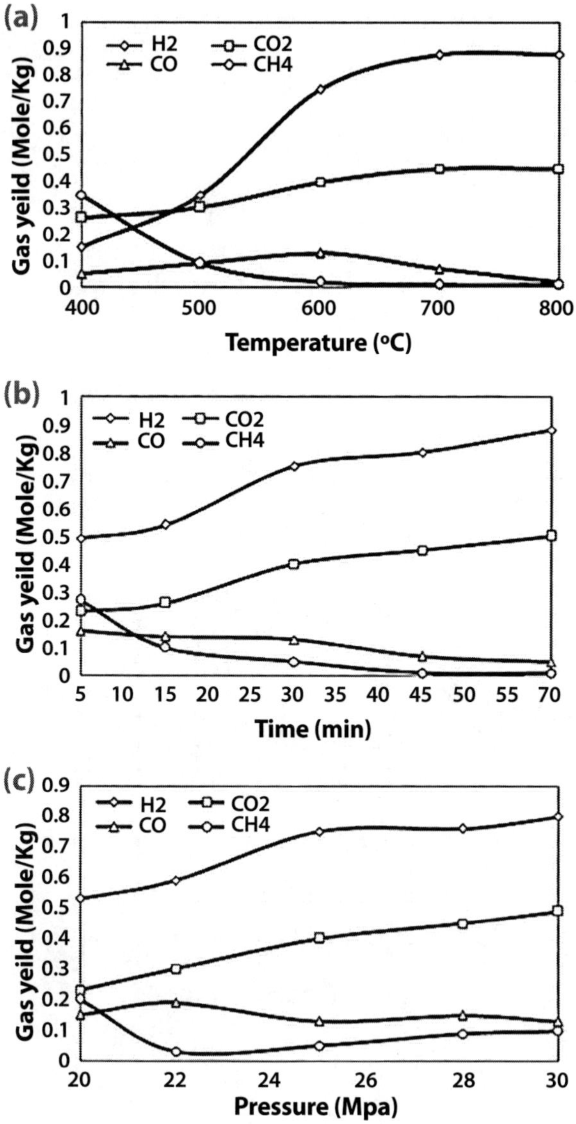

FIGURE 2: a Gas yeild using different temperature at fixed 25 MPa pressure and time 30 min, b Gas yeild using differnet retension time at fixed 600 °C temperature and 25 Mpa pressure, c Gas yield using different pressure at fixed 600 °Ctemperature and time 30 min

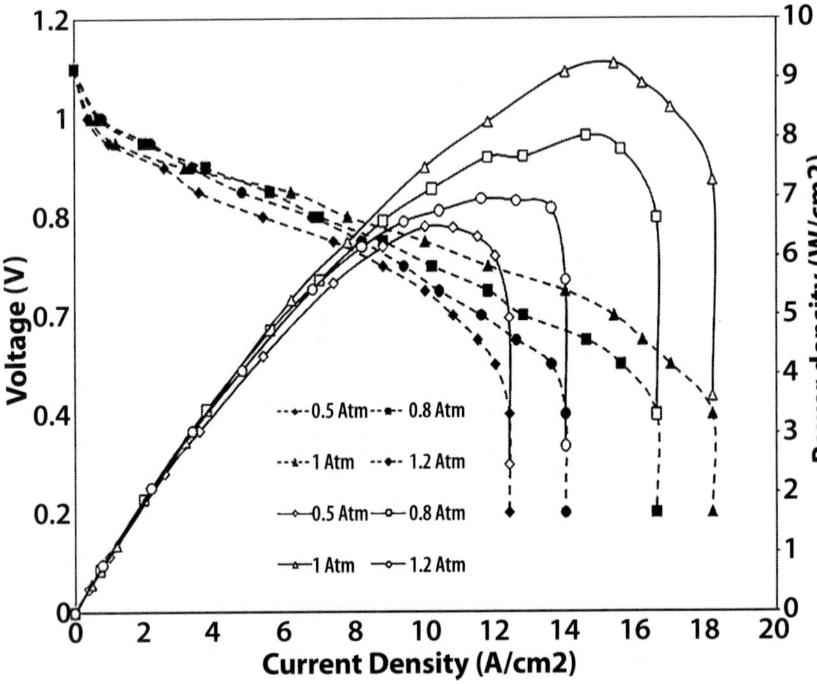

FIGURE 3: Cell performance at different partial pressure at 65 °C and 2 mol electrolyte concentration

4.3.4 EFFECT OF PRESSURE

The effect of pressure on equilibrium gas yields at 600 °C at 30 min using different pressure. At 25 MPa the H_2 and CO_2 were 75 and 40 % of yields, with increase in pressure; little changes were found, which are shown in Fig. 2c. Pressure shows a complex effect on biomass gasification in SCW. The properties of water, such as density, static dielectric constant and ion product increase with pressure. As a result, the ion reaction rate increases and free-radical reaction is restrained with an increase of pressure. Hydrolysis reaction plays a significant role in SCWG of biomass, which requires the presence of H^+ or OH^-. With increasing pressure, the ion product increases, therefore the hydrolysis rate also increase. Besides, high pressure favors water–gas shift reaction, but reduces decomposition reaction rate. But in the case of CH_4 and CO it was very less or negiable but with the increase in the pressure it increased slowly. The complex pressure effects can be used to fine tune the chemical composition of the solvent and control gas composition with yield [28]. Specifically, pressure has little or no influence on reaction rate, but it does affect solvent density. Density also has little effect on gasification efficiency above the critical point, but can have significant affects on gas fraction characteristics [29]. High pressures, and correspondingly higher densities, favor CH_4 production and inhibit H_2 production.

4.3.5 EFFECT OF PARTIAL PRESSURE OF HYDROGEN IN CELL PERFORMANCE PRESSURE

To investigate the effect of hydrogen partial pressure on the cell performance, the cell performance is studied with different hydrogen partial pressures of 0.5, 0.8, 1 and 1.2 atm, respectively. The partial pressure is adjusted by mixing argon. The cell performance with different partial pressures of hydrogen at a temperature of 65 °C is shown in Fig. 3. The power density rises with the increase of the hydrogen partial pressure. The increase in cell open circuit voltage will be somewhat less because of the greater gas solubility at increasing pressure which produces higher lost currents. When the partial pressure of hydrogen is higher than 0.8 atm, the cell can keep a high output performance as that of cell using pure hydrogen. The maximum power density of 9.24 W/cm^2 was obtained using pure hydrogen. The maximum power density only decreased by about 13 % when the partial pressure of hydrogen decreased from 1 atm to 0.8 and 1.2 atm. However, the cell performance decreases dramatically as the partial pressure of hydrogen decreases to 0.5 atm. The maximum power density is only 6.5 W/cm^2 at a partial pressure of hydrogen of 0.2 atm.

The OCV can be calculated from the Nernst equation:

$$E = E^0 + \left(\frac{RT}{2F}\right)\ln\left(\frac{P_{H_2}}{P_{H_2O}}\right) + \left(\frac{RT}{2F}\right)\ln(P_{O_2}^{1/2}) \quad (1)$$

where E° is the open circuit voltage (OCV) at standard pressure. F is the Faraday constant and R is the gas constant. T is the absolute temperature. P_{O2} is the partial pressure at the cathode. And P_{H2} and P_{H2O} are the partial pressures of hydrogen and vapour at the anode. It can be found that the OCV is dependent on cell temperature, hydrogen and water concentration in the fuel gas (anode) and oxygen in the cathode.

The above equation can be changed to:

$$E = E^0 + \left(\frac{RT}{2F}\right)\ln\left(\frac{P_{O_2}^{1/2}}{P_{H_2O}}\right) + \left(\frac{RT}{2F}\right)\ln(P_{H_2}) \quad (2)$$

If the hydrogen pressure changes from P_1 to P_2 and the partial pressures of P_{H2O} and P_{O2} keep constant, there will be a change of voltage at 65 °C:

$$\Delta V = \frac{RT}{2F}\ln(P_2) - \frac{RT}{2F}\ln(P_1) = \frac{RT}{2F}\ln\left(\frac{P_2}{P_1}\right) = 0.053\ln\left(\frac{P_2}{P_1}\right)V \quad (3)$$

These experimental data are in good agreement with the difference of reversible cell voltage calculated based on Nernst equation.

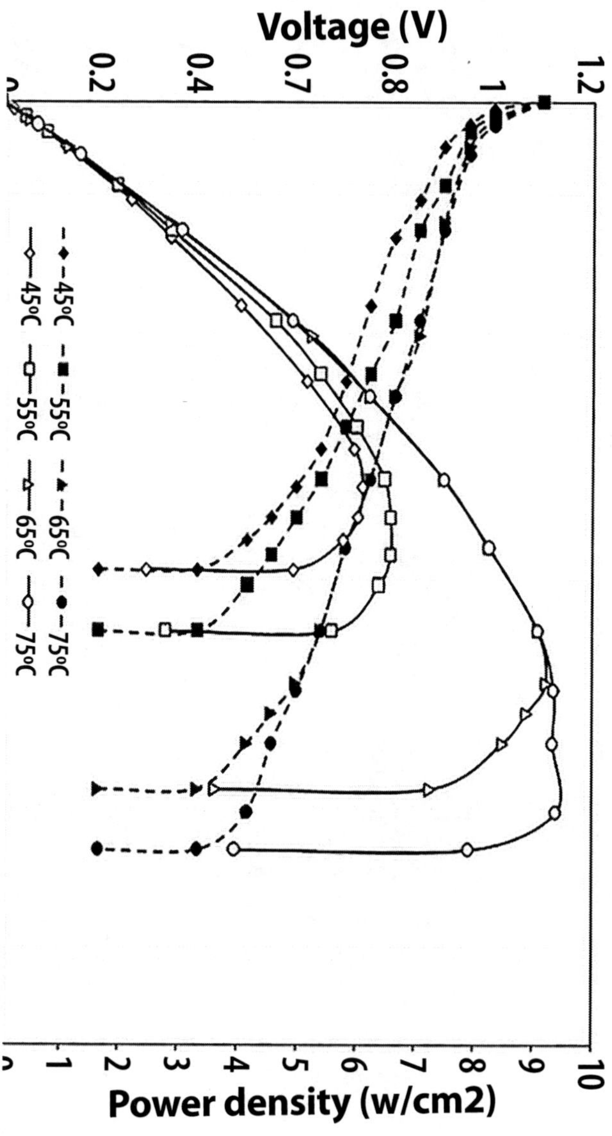

FIGURE 4: Cell performance using different temperature at 1 atm. Pressure and 2 mol electrolyte concentration

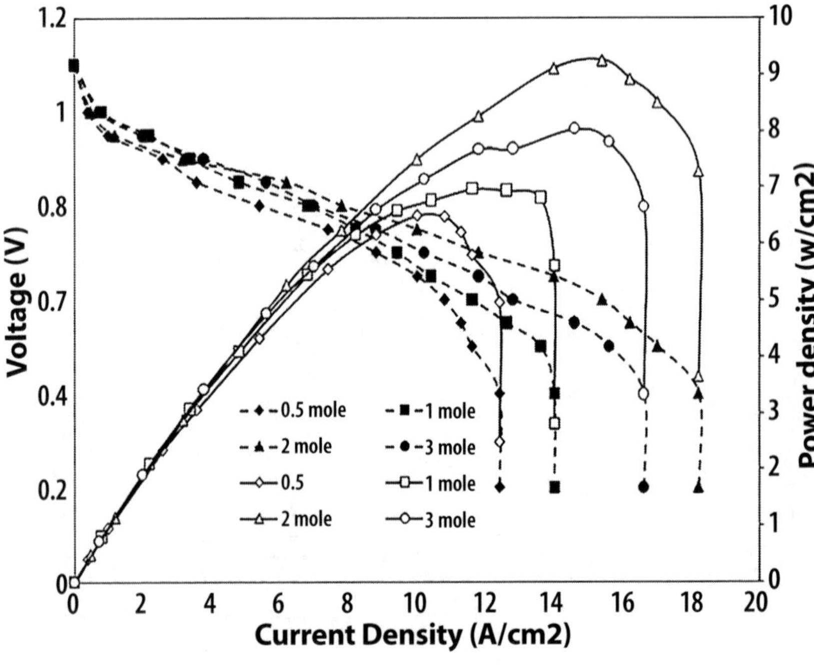

FIGURE 5: Cell performance using different fuel concentration at 65 °C temperature and 1 atm pressure

4.3.6 EFFECT OF TEMPERATURE IN CELL PERFORMANCE

In the cell, reaction process becomes faster when the electrolyte is warm rather than cold. So, the temperature plays an important role to develop the voltage across terminal. Figure 4 shows the current–voltage relationship for using different temperature at 1 atm and 2 M of electrolyte solution was fed to the alkaline fuel cell. It is seen that the cell performance increases with the increase in temperature because of decrease in the activation and concentration over-potentials [30]. In addition, mass transport limitations are reduced at higher temperatures. The overall result is an improvement in cell performance or in other words the conductivity of KOH solutions is relatively high at low temperatures. For instance an alkaline fuel cell designed to operate at 75 °C will reduce to only half power level when its operating temperature is reduced to room temperature. The maximum power density 9.36, 9.24, 6.6, and 6.12 W/cm^2 was obtained when temperature is 75, 65, 55, and 45 °C. This result was found similar to author, who used methanol and ethanol fuel at 25, 45 and 65 °C. The performance increases with the increase in temperature because of decrease in activation over potential concentration and mobility at higher temperature [31].

4.3.7 EFFECT OF ELECTROLYTE CONCENTRATION IN CELL PERFORMANCE

A higher current flow (amperage) through the cell means it will be passing more electrons through it at any given time. This means a faster rate of reduction at the cathode and a faster rate of oxidation at the anode. This corresponds to a greater number of moles of the product. The amount of current that passes depends on the concentration of the electrolyte; it shows different value in different concentration of electrolyte used. Figure 5 shows that the cell voltage increases with the increase in KOH concentration from 1 to 2 M for a particular load and then it decreases with further increase in KOH concentration. It is well known that the initial and final voltage losses with an increase in current consumption and are attributed to activation and concentration over-potentials, whereas the over-potential in the flattened portion of the curve is due to ohmic loss [32]. It is apparent from Fig. 5 that the increase in KOH concentration has minimum effect on activation over-potential while the concentration over-potential first decreases and then increases with the increase in KOH concentration. The concentration polarization increases at a higher KOH concentration because of less availability of hydrogen at the anode. On the other hand, the lowering of the KOH concentration increases the ionic conductivity of the medium or decreases the ohmic loss. The cell performance is maximum at 2 M of electrolyte concentration obtained at 9.24 W/cm^2, and lowest at 0.5 M was 6.48 W/cm^2. While further increase in electrolyte beyond 2 M, it was found the cell performance decreases.

4.4 CONCLUSION

It concludes that the contents of hydrogen and carbon dioxide vary with different operation conditions and obtained 88.8 % of hydrogen and approximate 45 % of carbon dioxide at temperature(600 °C), pressure(25 MPa), and residence time(60 min). Although supercritical water gasification of wet biomass seems promising for the production of hydrogen rich gas, it should be noticed that a high concentration of biomass is necessary to reach commercial goals. From the experiment we find the maximum power density 9.24 W/cm^2 was obtained at 75 °C (Temperature) 2 M (Electrolyte concentration) and 1 atm (Pressure). The development of hydrogen and fuel-cell technologies is set to play a central role in addressing growing concerns over carbon emissions and climate change as well as the future availability and security of energy supply. Hydrogen can be generated from biomass, but this technology urgently needs further development. It is believed that in the future, biomass can become an important sustainable source of hydrogen. Due to its environmental merits, the share of hydrogen from biomass in the automotive fuel market will grow fast in the next decade. Gasification of biomass has been identified as a possible system for producing renewable hydrogen, which is beneficial to exploit biomass resources, to develop a highly efficient clean way for large-scale hydrogen production, and has less dependence on insecure fossil energy sources. Steam reforming of natural gas and gasification of biomass will become the dominant technologies by the end of the 21st century.

REFERENCES

1. Dincer, I.: Technical, environmental and exergetic aspects of hydrogen energy systems. Int. J. Hydrog. Energy. 27, 265–285 (2007)
2. Arena, U.: Process and technological aspects of municipal solid waste gasification. A review. Waste. Manag. 32, 625–639 (2012)
3. Baratieri, M., Baggio, P., Fiori, L., Grigiante, M.: Biomass as an energy source: thermodynamic constraints on the performance of the conversion process. Biotechnol. 99, 7063–7073 (2008)
4. Franco, A., Giannini, N.: Perspectives for the use of biomass as fuel in combined cycle power plants. Int. J. Therm. Sci. 44, 163–168 (2005)
5. Anon, C.: Reducing greenhouse gas emissions and electrical power costs. Biocycle. 45(10), 35–36 (2004)
6. Appleby, A.J., Foulkes, F.R.: Fuel cell handbook. Van Nostrand Reinhold, New York (1993)
7. Dutton, AG.: Hydrogen energy technology. Tyndall Working Paper TWP17, Tyndall Centre for Climate Change. http://www.tyndall.ac.uk/publications/working_ papers/wp17.pdf (2008). Accessed 15 May 2008
8. Crabtree, G.W., Dresselhaus, M.S., Buchanan, M.V.: The hydrogen economy. Phy. Today. 57, 39–44 (2004)
9. Novochinskii, I., Ma, X., Song, C., Lambert, J., Shore, L., Farrauto, R.: A ZnO-based sulfur trap for H2S removal from reformate of hydrocarbons for fuel cell applications. Proceedings of Topical Conference on Fuel Cell Technology. AIChE Spring National Meeting, New Orleans, vol. 11–14, pp. 98–105. (2002)
10. Turner, J.A.: Sustainable hydrogen production. Inform. 305, 971–974 (2004)
11. Unal, H., Alibas, K.: Agricultural residues as biomass energy. Energy. Sources. Part. B. 2, 123–140 (2007)
12. Boerrigter, H., Rauch, R.: Review of applications of gases from biomass gasification. In: Knoef, HAM. (ed.) Proceedings of the handbook biomass gasification. The Netherlands: Biomass Technology Group (BTG). pp. 211–230 (2005)
13. Barreto, L., Makihira, A., Riahi, K.: Medium and long-term demand and supply prospects for fuel cells: the hydrogen economy and perspectives for the 21st century. International Institute for Applied Systems Analysis, Laxenburg (2002)
14. Sealock, L.J.J., Elliott, D.C., Baker, E.G., Fassbender, A.G., Silva, L.J.: Hydrogen production by supercritical water gasification. Ind. Eng. Chem. Res. 35, 4111–4120 (1996)
15. Clean Renewable Fuel from the Plasma Gasification of Waste, http://www.waste-management-world.com (2011). Accessed 16 May 2011

16. Elliott, D.C., Hart, T.R., Neuenschwander, G.G.: Chemical processing in high pressure aqueous environments: improved catalysts for hydrothermal gasification. Ind. Eng. Chem. Res. 45(11), 3776–3781 (2006)
17. Soria, J.A., McDonald, A.G., Shook, S.R.: Wood solubilization and depolymerization using supercritical methanol. Part 1: process optimization and analysis of methanol insoluble components (Bio-Char). Holzforschung 14(4), 402–408 (2008)
18. Gloyna, E.F., Li, L.: Supercritical water oxidation: an engineering update. Waste. Manag. 14, 379–394 (1993)
19. D'Jesus, P., Boukis, N., Kraushaar-Czarnetzki, B., Dinjus, E.: Gasification of cornand clover grass in supercritical water. Fuel 85, 1032–1038 (2006)
20. Elliott, D.C.: Catalytic hydrothermal gasification of biomass. Bioprod. Biorefining. 2(3), 254–265 (2008)
21. Susanti, R.F., Veriansyah, B., Kim, J.D., Kim, J., Lee, Y.W.: Continuous supercritical water gasification of isooctane: a promising reactor design. Int. J. Hydrog. Energy. 35:51957–1970 (2010)
22. Cao, C., Guo, L., Chen, Y., Guo, S., Lu, Y.: Hydrogen production from supercritical water gasification of alkaline wheat straw pulping black liquor in continuous flow system. Int. J. Hydrog. Energy. 36(21), 13528–13535 (2011)
23. Antal, M.J., Allen, S.G., Schulman, D., Xu, X., Divilio, R.J.: Biomass gasification in supercritical water. Ind. Eng. Chem. Res. 39(11), 4040–4053 (2000)
24. Xu, L., Brilman, D.W.F., Withag, J.A.M., Brem, G., Kersten, S.: Assessment of a dry and a wet route for the production of biofuels from microalgae: energy balance analysis. Bioresour. Technol. 102(8), 5113–5122 (2011)
25. Xu, X., Matsumura, Y., Stenberg, J., Antal, M.J.: Carbon-catalyzed gasification of organic feed stocks in supercritical water. Ind. Eng. Chem. Res. 35(8), 2522–2530 (1996)
26. Savage, P.E.: A perspective on catalysis in sub- and supercritical water. J. Supercrit. Fluids. 47(3), 407–414 (2009)
27. Du, X., Zhang, R., Gan, Z., Bi, J.: Treatment of high strength coking wastewater by supercritical water oxidation. Fuel. (2010). doi:10.1016/j.fuel.2010.09.018
28. Afif, E., Azadi, P., Farnood, R.: Catalytic hydrothermal gasification of activated sludge. Appl. Catal. B. Environ. 105, 136–143 (2011)
29. Brunner, G.: Near and supercritical water. Part II: oxidative processes. J. Supercrit. Fluids. 47(3):382–390 (2009)
30. Pramanik, H., Basu, S.: Modeling and experimental validation of overpotentials of a direct ethanol fuel cell. Chem. Eng. Process. 49(7), 635–642 (2010)
31. Gaurav, D., Verma, A., Sharma, D., Basu, S.: Development direct alcohol alkaline fuel cell stack. Fuel. Cell. 10(4), 591–596 (2010)
32. Koscher, GA., Kordesch, K.: Alkaline methanol/air power devices, in: Handbook of fuel cells—fundamentals, technology and applications. In: Vielstich, W., Gasteiger, H.A., Lamm, A. (eds.), John Wiley, 4:1125–1129 (2003)

CHAPTER 5

Landfill Minimization and Material Recovery via Waste Gasification in a New Waste Management Scheme

N. TANIGAKI, Y. ISHIDA, AND M. OSADA

5.1 INTRODUCTION

Waste-to-Energy is attracting great interest in many countries, such as in Japan and in Europe. In Japan, though the volume reduction of waste has historically been the first priority in waste management, energy from waste is also becoming a great interest these days. In Europe, due to the EU landfill directive, a large number of Waste-to-Energy plants have been established to reduce landfill amounts and recover energy (European Union, 1999). On the other hand, material recovery from urban sites, which indicates the recovery of valuable materials and noble metals such as gold, copper and ferrous materials in waste including municipal solid waste (MSW), MSW bottom ash, incombustibles from recycling centers, boiler

Tanigaki N, Ishida Y, and Osada M. "Landfill Minimization and Material Recovery Via Waste Gasification in a New Waste Management Scheme." Proceedings Venice 2014, Fifth International Symposium on Energy from Biomass and Waste, *San Servolo, Venice, Italy; 17–20 November 2014.*

and fly ash from MSW incinerators, are also attracting considerable major interest as well as energy recovery from waste (De Boom et al., 2011).

There is also concern regarding material recovery and recycling of bottom ash, though it has long been attracting great interest. Due to intermediate temperature treatment of incineration technology, the bottom ash discharged sometimes contains harmful heavy metals such as lead, zinc and mercury. Therefore, the leaching behavior of heavy metals needs to be carefully monitored. A large body of research has been reporting on the leaching behavior of heavy metals in bottom ash and its environmental impacts (Hyks et al., 2009; Birgisdorttir et al., 2007).

From the viewpoint of regulations, bottom ash recycling and utilization differ from one country to another. In some countries, it can be recycled as secondary materials after post-treatment such as aging. Aging is a long-term maturing process and harmful heavy metals in bottom ash are stabilized for two or three months. The stabilized bottom ash is recycled, mainly in road construction or as cover soil for final landfill sites. On the other hand, in some countries, such as Japan, Austria and Finland (CEWEP, 2010; ISWA, 2006), the bottom ash discharged from waste incineration facilities is not allowed to be recycled. It is mainly transported and disposed of in a landfill site. In these countries, further research or an alternative to bottom ash stabilization and material recovery from waste are needed.

A waste gasification and melting technology has the possibility of simultaneously addressing both material recovery from waste and landfill minimization. Gasification of MSW and biomass as an energy recovery method such as fixed-bed, fluidized bed, entrained or plasma gasification, has been widely researched all over the world (Arena et al., 2010; Arena and Di Gregorio, 2013(a); ; Mastellone et al., 2008, 2010;; Aigner et al., 2011; Pinto et al., 2008; Taylor et al., 2013; Willis et al., 2010). Waste gasification is also being researched in many countries, but there are few proven or commercial technologies. Furthermore, in many cases, limited feedstock, such as high-quality refused derived fuel (RDF), is processed in demonstration or commercial gasification plants. In addition to the above-mentioned issues, there is no significant improvement of residues discharged from developing gasification technologies. In Japan, a lot of gasification plants for MSW, including gasification and melting processes, are under commercial operation. The Direct Melting System (DMS) is a shaft-furnace type gasification and melting process and is classified as an atmospheric moving bed gasifier. The DMS has been in commercial operation since 1979 and has been operated at more than 40 plants in Japan and South Korea (Tanigaki et al., 2012, 2013a; Osada et al., 2008, 2012; Manako et al., 2007). This technology can process various kinds of waste, such as asbestos, bottom ash, clinical waste, sewage sludge or automobile shredder residues (ASR) with MSW, and can recover both energy and material from waste in one process. In this study, this material recovery from various kinds of waste via gasification and melting process (the DMS technology) is defined as co-gasification, due to the contemporary feeding of different wastes. The molten materials produced, which can be separated in inert slag and metal, can be utilized as recyclables without any further post-treatment such as aging. As a positive consequence of this waste flexibility and material recovery from waste, this technology can offer a new waste management scheme. This new waste management scheme can process MSW with other waste such as incombustibles or bottom ash from other incineration plants to maximize material recovery from waste and minimize final landfill amounts.

The objective of this study is to clarify the possibilities of the new waste management scheme using co-gasification of MSW with bottom ash and incombustible residues. Firstly, the operating data of co-gasification is evaluated using a commercial waste gasification and melting technology (DMS). The quality of slag and metal produced are also investigated to evaluate the potential for material recovery from waste. Secondly, a case study in two different waste management schemes is conducted and compared with other conventional systems: an incineration without a bottom ash melting system and with a bottom ash melting system. Lastly, sensitivity analyses are also conducted with variation of the landfill gate fee, power price and inert contents of waste to be processed. Some operation performances of the DMS have already been reported (Osada et al., 2008, 2012; Manako et al., 2007; Tanigaki et al., 2012, 2013a). However, this is the first report which describes the main economic aspects of co-gasification of MSW and other waste.

5.2 METHODOLOGY

5.2.1 PROCESS DESCRIPTION

The process is described in previous papers (Tanigaki et al., 2012, 2013a). The plant mainly consists of an MSW charging system, a gasifier, a combustion chamber, a boiler and a flue gas cleaning system.

One of the advantages of the DMS process is that no pretreatment of waste is required, which differs from other gasification technologies such as a fluidized bed gasifier. The maximum waste size to be processed is approximately 800 mm. Waste is directly charged into a gasifier from the top, together with coke and limestone which function as a reducing agent and a viscosity regulator, respectively. Due to the addition of the limestone, the viscosity of the melt is adjusted properly and the melt is discharged smoothly from the bottom of the furnace without any clogging. Oxygen-enriched air is blown at the bottom of the furnace via tuyere. The gasifier consists of three main parts: a drying and preheating zone, a thermal decomposition zone and a combustion and melting zone. waste is gradually dried and preheated in the upper section (the drying and preheating zone). Combustible waste is thermally decomposed in the second zone and syngas is discharged from the top of the gasifier. The syngas, which mainly contains carbon monoxide, carbon dioxide, hydrogen, methane, hydrocarbons and nitrogen, is transferred to the combustion chamber in the downstream of the gasifier and then completely burned. Incombustibles descend to the combustion and melting zone (1,000–1,800 °C) at the bottom and are melted with the heat generated by coke burning. Gasification reactions such as water-gas-shift, water-gas and boudouard reactions mainly take place in this zone of the gasifier. Molten materials are intermittently discharged from a tap hole, quenched with water, and magnetically separated into slag and metal. Due to high-temperature and reducing atmosphere generated by coke burning at the bottom of the gasifier, toxic heavy metals are volatilized and few toxic heavy metals remain in molten materials, which can be widely recycled. Because of this, the DMS can maximize material recovery from waste without any further pre- and post-treatment such as drying, crashing or aging.

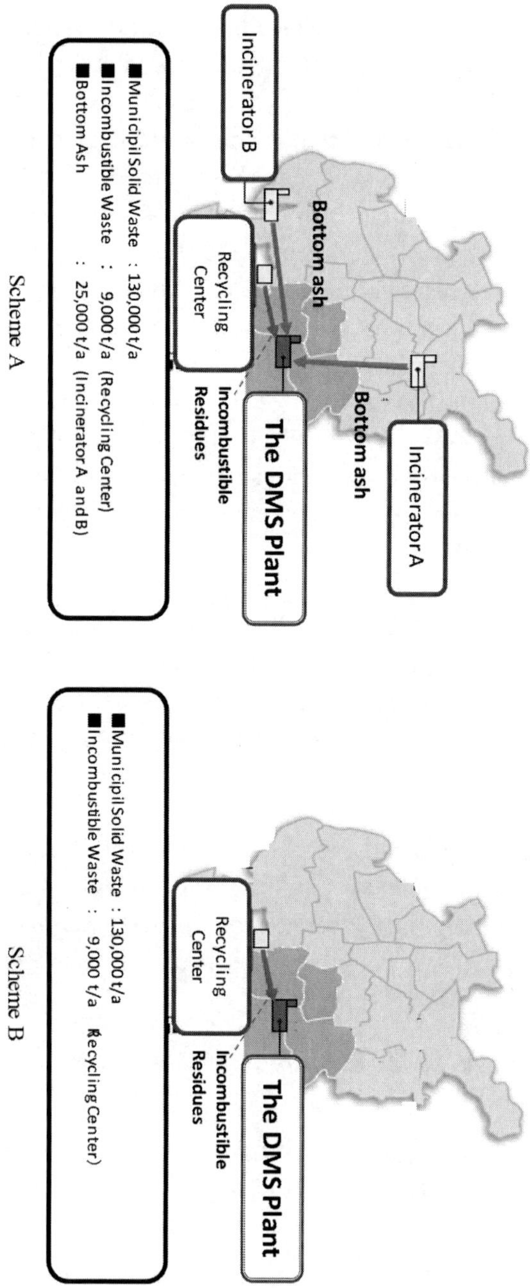

FIGURE 1: Conceptual diagram of the new waste management

The combustible dust discharged from the gasification and melting furnace is captured by the cyclone installed in the downstream of the furnace. The captured char is injected into the furnace via tuyere. This system has three major advantages (Manako et al., 2007). Firstly, the coke amount is reduced because the injected combustible dust reacts with oxygen-rich blown air as a coke substitute. Secondly, the combustion condition in the combustion chamber is improved because of gas combustion with little dust. Lastly, dust is captured by the cyclone and this reduces fly ash amount in the downstream. In addition to this technology, an air pre-heating technology for combustible dust injection can reduce the coke amount dramatically (Tanigaki et al.,2013b). This technology has improved the reaction kinetics of the char injected into the gasifier. As a result, the coke consumption is reduced between 2% and 4% with the application of these technologies. Biomass coke application with these developed technologies can also reduce the environmental impacts such as global warming (Tanigaki et al., 2012).

The syngas produced in the gasifier is combusted in the combustion chamber by air. Heat is recovered by a steam boiler and power is generated by a steam turbine. Flue gas cleaning system is similar to that of incineration technologies (Tanigaki et al. 2013a). That generally consists of a quencher, a baghouse, an ID-fan and a selective catalytic reduction (SCR). Flue gas is cooled to approximately 150 °C in the quencher. $Ca(OH)_2$ injection is applied for desulfurization and is installed at the upstream of the baghouse. The re-heater and SCR are also applied for nitrogen oxides (NOx) reduction. A selective non-catalytic reduction (SNCR) system can be also applied in the DMS technology. Fly ash captured at the boiler and the baghouse are transferred to a fly ash treatment system.

5.2.2 THE NEW WASTE MANAGEMENT SCHEME

The new waste management scheme was developed using the co-gasification system. This system processes the bottom ash from other incineration plants and/or incombustible residues from a recycling center with MSW using the DMS technology and recycles the inert materials as slag and metal. This indicates that ash contents in the waste can be converted into inert materials and recyclables, and landfill amount can be minimized. In this study, two schemes with different boundary conditions are evaluated. A conceptual diagram of the new waste management is shown in Figure 1.

The first scheme (scheme A) is co-gasification of bottom ash and incombustible residues with MSW containing low inert materials. The municipality has already been operating three waste to energy plants and one of them is considered to be replaced by a co-gasification system. In this scheme, the new plant processes 130,000 t/annum of MSW discharged from the area highlighted by dark green in Figure 1. In addition to MSW, the co-gasification system processes the bottom ash from other two incineration plants and incombustibles from a recycling center. The annual bottom ash generation from other incineration plants and incombustibles from a recycling center are set as 25,000 t/annum and 9,000 t/annum. The inert contents and net calorific value (NCV) of MSW are set as $11.8\%_{a.r.}$ and 9.0 $MJ/kg_{a.r.}$, respectively. These boundary conditions are based on our previous study (Tanigaki et al., 2012).

The other scheme (scheme B) is based on MSW with high inert materials, which is different from that in scheme A. In this scheme, the municipality has an existing recycling center and employs a new waste-to-energy plant using the co-gasification system to process MSW discharged from highlighted area in Figure 1 and incombustibles from an existing recycling center and to minimize final landfill amount. The total annual amounts of MSW and incombustibles are set as the same as in scheme A. The inert contents and the NCV of MSW are set as $22\%_{a.r.}$ and 9.0 $MJ/kg_{a.r.}$, respectively. This waste composition is in the range of that in Europe (European Commission, 2006). This waste composition is different from that in Japan (scheme A). However, this difference of waste composition between in Japan and in Europe doesn't affect the operating conditions of the DMS technology such as coke consumption significantly. One of the important factors for the operating conditions of the DMS technology is the moisture contents of the waste processed. When the moisture contents of the waste is reduced, the coke consumption can be reduced, because higher temperature can be kept with less energy for drying and pre-heating of the waste charged. In addition, the inert contents of the waste are similar to that of our previous study (Tanigaki et al., 2012). Because of these reasons, the difference between Japanese and European waste doesn't have a big influence on the operating conditions of the DMS technology.

For both schemes, the co-gasification systems in the new waste management scheme are compared with two alternatives: incineration without the bottom ash melting system and incineration with a bottom ash melting system. The capacity of the incineration plant is set as 130,000 t/annum. In the alternative cases, the new plant processes only MSW generated from the highlighted region in Figure 1. No bottom ash from other two incinerators is co-processed with MSW in the new plant. The bottom ash discharged from the new plant and other two incineration plants is landfilled or stabilized with the bottom ash melting system. In both the alternative cases, incombustibles from an existing recycling center are directly transferred to landfill sites.

TABLE 1: Boundary and pre-conditions

		Scheme A			Scheme B		
		Case 1	Case 2	Case 3	Case 4	Case 5	Case 6
		DMS Co-Gasification	Incineration		DMS Co-Gasification	Incineration	
			With bottom ash melting	Without bottom ash melting		With bottom ash melting	Without bottom ash melting
PROCESSED	MSW	X	X	X	X	X	X
	Bottom ash from other plants	X	landfilled	vitrified	-	-	-
	Incombustibles	X	landfilled	landfilled	X	landfilled	landfilled
	Bottom ash from new plant	-	landfilled	-	-	landfilled	-
OUTPUT	Slag	Recycled	-	Partly Recycled	Recycled	-	Partly Recycled
	Incompatibility Rates	0%	-	14%	0%	-	14%
	Metal	Recycled	-	-	Recycled	-	-
	Fly ash	landfilled	landfilled	landfilled	landfilled	landfilled	landfilled

The slag and metal produced from the DMS are recycled completely, with reference to recycling conditions in Japan. Based on this recycling condition of slag in Japan, the incompatibility rates of slag and metal in the DMS were assumed to be 0%. There is less information on the molten material recovery via ash melting technologies. The Japanese Organization of Research Institute of Innovative Technology for the Earth (RITE) (RITE, 2010) has reported that molten materials in the ash melting systems in Tokyo were generally recycled with incompatibility rates of 28%. This indicates that 72% of molten materials in the ash melting process can be recycled and 28% is landfilled. In this study, we assume incompatibility rates of 14%, considering technical improvements. The incineration bottom ash is not allowed to be recycled and is transferred to final landfill sites. All the fly ash is also landfilled and no recovery of heavy metals in the fly ash is considered in this study. The fly ash amounts of the DMS and the incineration are set as 3.3% and 3.0%, respectively (Tanigaki et al., 2012). The detailed boundary conditions and preconditions of this study are shown in Table 1.

5.3 RESULTS AND DISCUSSION

5.3.1 OPERATION DATA OF THE CO-GASIFICATION SYSTEM

5.3.1.1. OPERATION DATA OF THE CO-GASIFICATION SYSTEM

Operation data of the co-gasification system are shown in Table 2. This operating data is taken in one of the DMS commercial plant in Japan (Tanigaki et al., 2012). Slag, metal produced and fly ash discharged in the plant were 173 kg/t-waste, 27 kg/t-waste and 33 kg/t-waste, respectively. Fly ash is defined as a mixture of raw dust from the system and de-acidification agents (slaked lime). The syngas temperature was 517 °C, which is much higher than the condensation temperature of tar. This indicates that there is no clogging trouble caused by tar in the DMS. The operating results such as total waste throughput, coke consumption, slag, metal produced and fly ash discharged can also be used for the calculation of material balance in scheme B (3.2.1.), because the inert contents of the waste processed is similar to those of European waste.

TABLE 2: Operation results (Tanigaki et al., 2012)

	Unit	Values
Total waste thoughput	t/day	251.8
MSW	t/day	200.1
Bottom ash from existing plants	t/day	40
Incombustibles from a recycling center	t/day	11.7
Coke	kg/t-MSW	44.9
Top gas temperature of the furnace	°C	517
Slag produced	kg/t-MSW	173
Metal produced	kg/t-MSW	27
Fly ash	kg/t-MSW	33

TABLE 3: Compositions of slag, metal and fly ash (Tanigaki et al., 2012)

	Slag	Metal	Fly ash
Unit	$\%_{db}$	$\%_{db}$	$\%_{db}$
SiO_2	38	2.1	6.9
CaO	32	0.2	25
Al_2O_3	13	0.085	4.0
MgO	2.0	0.021	1.0
Pb	0.0039	0.043	1.1
Zn	0.066	0.11	5.1
Na	2.4	0.076	7.1
K	0.80	0.034	7.8
Cl	0.0	0.0	15
S	0.21	0.14	2.9
Cu	0.052	8.3	0.28
Fe	2.3	81	0.55

5.3.1.2 SLAG AND METAL COMPOSITIONS

Table 3 shows an example of the compositions of slag, metal produced and fly ash discharged from the co-gasification system with the DMS technology (Plant A) shown in our previous paper (Tanigaki et al., 2012). Slag mainly consists of SiO_2, CaO and Al_2O_3. Lead and zinc concentrations are significantly lower, at $0.0039\%_{db}$ and $0.066\%_{db}$, respectively. Iron is a main component in the metal, at $81\%_{db}$. There is also a high distribution of copper in the metal. The main components of the fly ash are CaO and chlorine, the concentrations of which are $25\%_{db}$ and $15\%_{db}$, respectively.

Table 4 shows the results of the slag leaching tests (in compliance with JIS K0058). These results show that the slag produced contains almost no harmful heavy metals such as lead. The value of lead resulting from the slag leaching tests was less than 0.005 mg/L, which was remarkably lower than the target reference (less than 0.01 mg/L). In the DMS, heavy metals are distributed in each fraction following thermodynamic behaviors. According to a thermodynamics calculation, lead and zinc in the form of sulfide or oxide are moved to slag and Osada et al. (2009 and 2010) showed that a high temperature reducing atmosphere is effective in preventing the formation of lead and zinc sulfide or oxide. Because of these reactions, the slag contains little lead and zinc.

The results of an acid-extractable test are also shown in Table 4. In the acid-extractable test, 1 mol/L hydrochloric acid is used as an extraction solvent. The mixing ratio of the sample to the solvent is set as 3%-vol. The sample with an acid solvent is shaken at 200 rpm for two hours. After shaking, heavy metal concentrations in the solvent are detected (acid-extractable test). The value of lead in the acid-extractable test (in compliance with JIS K0058) was 18.0 mg/kg, which is far lower than the target reference (less than 150 mg/kg). Likewise, the values of Cd and other heavy metals in the slag leaching test and acid-extractable test were remarkably lower than the target reference.

The above findings clarify that the molten slag resulting from processing using the DMS is different from that of bottom ash, and contains few harmful heavy metals such as lead, and can be used as interlocking block aggregate, concrete aggregate, or civil engineering material without additional post-treatment. This implies that that the DMS technology can allow a reduction up to approximately 65% of the amount of waste to be landfilled, when compared to the conventional combustion-based technology (Arena and Di Gregorio, 2013(b)). This finding is a crucial difference between the DMS technology and conventional combustion-based technologies and is an important parameter for evaluating this case study

TABLE 4: Compositions of slag, metal and fly ash (Tanigaki et al., 2012)

	Leaching test		Acid-extraction test	
	Target reference	Measured	Target reference	Measured
Cd	< 0.01 mg/L	< 0.001	< 150 mg/kg	< 5
Pb	< 0.01 mg/L	< 0.005	< 150 mg/kg	18
Cr^{6+}	< 0.05 mg/L	< 0.02	< 250 mg/kg	< 5
As	< 0.01 mg/L	< 0.001	< 150 mg/kg	< 5
T-Hg	< 0.0005 mg/L	< 0.0005	< 15 mg/kg	< 0.05
Se	< 0.01 mg/L	< 0.001	< 150 mg/kg	< 5
CN	-	-	< 50 mg/kg	< 1
F	-	-	< 4000 mg/kg	172
B	-	-	< 4000 mg/kg	260
Metal- Fe	-	-	< 1.0%	0.18

**1 Both results are conducted six times.*

5.3.2 CASE STUDY OF THE CO-GASIFICATION SYSTEM

5.3.2.1 MATERIAL BALANCE

The material balance in this study is shown in Table 5. The "INPUT" shows all the materials to two existing incinerators and the new facility (the DMS or an incinerator). The "PROCESSED" shows the waste to be processed in the new plant. In scheme A, two existing incinerators process MSW with an amount of 265,000 t/annum and discharge the bottom ash with a total amount of 25,000 t/annum. As explained above, the processing amount of MSW in the new plant is set as 130,000 t/annum. In case

1, the total waste processing amount of the co-gasification system is set as 164,000 t/annum, because the incombustibles and the bottom ash from other incinerators are processed together with MSW. In the cases 2 and 3, incombustibles and bottom ash are not processed in the new incineration plant. In case 4 of scheme B, the co-gasification system processes MSW and incombustibles with a total amount of 139,000 t/annum. Cases 5 and 6 process only MSW and incombustibles are disposed of.

TABLE 5: Material balance of the case study

			Scheme A			Scheme B		
			Case 1	Case 2	Case 3	Case 4	Case 5	Case 6
		unit		Incineration			Incineration	
			DMS	Without Ash Melting	With Ash melting	DMS	Without Ash Melting	With Ash melting
INPUT	MSW for new plant	t/a	130,000	130,000	130,000	130,000	130,000	130,000
	Incombustibles	t/a	9,000	9,000	9,000	9,000	9,000	9,000
	MSW for other two incinerators	t/a	265,000	265,000	265,000	-	-	-
PROCESSED	MSW	t/a	130,000	130,000	130,000	130,000	130,000	130,000
	Incombustibles	t/a	9,000	-	-	9,000	-	-
	Bottom ash from other incinerators	t/a	25,000	-	-	-	-	-
	Total waste amount for the new plant	t/a	164,000	130,000	130,000	139,000	130,000	130,000
OUTPUT	Slag*1	t/a	28,372		32,044	24,047		21,272
	Metal	t/a	4,428			3,753		
	Imcompatible slag*1	t/a	0	-	5,216	0	-	3,463
	Bottom ash	t/a	-	12,260	-	-	24,734	-
	Bottom ash from other incinerators	t/a	-	25,000	25,000	-	-	-
	Incombustibles	t/a	-	9,000	9,000	-	9,000	9,000
	Fly ash	t/a	5,412	3,900	3,900	4,587	3,900	3,900
	Final landfill amount	t/a	5,412	50,160	18,116	4,587	37,634	16,363

**1: Slag of cases 3 and 6 was calculated by the incompatibility ratio of 14% and can be recycled.*

TABLE 6: Parameter list

	Utility	Unit	Case 1, 4	Case 2,5	Case 3,6	Reference
			Incineration			
			DMS	Without Ash Melting	With Ash melting	
Gross power generation		%	23			-
Power consumption						
	Plant	kWh/t	170	100		Assumption
	Ash melting	kWh/t-residue	-	-	1272	RITE, 2010
Utilities						
	Coke	Euro/t	220	-		Assumption
	Natural gas	Euro/GJ	10.6			Eurostat (EU 27 Countries Average)
		MJ/m^3	38.3			-
	Other utilities	Euro/t	15	-	15	Assumption
	Power price	Euro/MWh	60			Assumption from CEWEP data [CEWEP, 2010]
Landfill						
	Gate fee	Euro/t-residue	118			Assumption based on CEWEP (2010), ISWA (2006) and Higuchi (2010) with 130 JPY/Euro
	Power consumption of vehicles	kWh/t-residue	64.5			Osada et al. 2012
	Diesel consumption for vehicles	L/t-residue	0.62			Osada et al. 2012
	Diesel oil price	Euro/L	1.4			www.energy.eu/fuel-prices/
Fly Ash Disposal		Euro/t-residue		250		Assumption

**The parameter list of cases 4, 5 and 6 are the same as cases 1, 2 and 3, respectively.*

The output data of the cases 1 to 6 is also shown in Table 5. Slag, metal and fly ash amounts in case 1 were calculated to be 28,372 t/annum, 4,428 t/annum and 5,412 t/annum, respectively, based on the operation data shown in Table 2. As mentioned above, the inert contents of the co-gasification plant (Plant A) is similar to that of scheme B. Therefore, the same operating data in Table 2 can be used for the case study in scheme B. The final landfill amount was only fly ash, because the slag and metal produced can be recycled completely as explained above in 3.1.2. In case 2, the total bottom ash amount in this scheme was calculated to be 37,260 t/annum. The fly ash discharged was 3,900 t/annum. The total final landfill amount was 50,160 t/annum. This value was nine times higher than that of case 1 because the bottom ash was not allowed to be recycled. Case 3 had a lower landfill amount than that in case 2. The ash melting system can significantly reduce the final amount to 18,116 t/annum. Due to the incompatibility ratio (14%), the final landfill amount of case 3 was slightly higher than that in case 1.

In case 4, the amounts of slag, metal discharged and fly ash were 24,047, 3,753 and 4,587 t/annum, respectively. In cases 5 and 6, bottom ash and fly ash discharged were 24,734 and 3,900 t/annum, respectively. Due to the ash melting system, the final landfill amount in case 6 was less than half of that in case 5.

5.3.2.2 ASSUMPTIONS

All the assumptions in this study are shown in Table 6. The gross power generation efficiency is set to be 23% for all the cases, based on a dry flue gas cleaning system (Tanigaki et. al, 2012), and the combined heat and power (CHP) is not considered in this study. The power consumptions of the DMS and an incineration plant are assumed to be 170 kWh/t-waste and 100 kWh/t-waste, respectively. Additional power consumption caused by oxygen generator in the DMS technology was taken into account. The power consumption of the ash melting system is set as 1,272 kWh/t-residue, based on RITE's report (2010). When the bottom ash discharged from MSW (scheme A) is processed by the ash melting system in the same facility, this assumption leads to a total power consumption of 220 kWh/t-waste in the facility. Oda et al. (2010) reported a similar value for the ash melting system. He reported that an incineration plant with an arc ash melting system in Japan (20 t/h with 9.3 MJ/kg of waste NCV) consumes 5 MW power in total. This indicates that the power consumption of the incineration plant with ash melting is 250 kWh/t-waste, which is similar to the assumption in this study.

Utilities prices assumed are also shown in Table 6. The price of coke used in the cases 1 and 4 is set as 220 Euro/t-coke. The Eurostat (2013) reported that the average natural gas price in the 27 EU countries average was 10.6 Euro/GJ. The additional costs for cases 1 and 3 are assumed to be 15 Euro /t-MSW. The power price is an important parameter. This study employs the power price of 60 Euro/MWh, based on previous reports (ISWA, 2006; CEWEP, 2010). Other costs such as operator's fee or maintenance costs are neglected in this study.

In this study, landfill costs are also assumed. The gate fee, power and diesel consumption of landfilling vehicles are also considered in evaluating these systems. The gate fee of a landfill site is an important parameter in this study. Higuchi (2006) reported that the CAPEX of a landfill site with the capacity of 144,000 m^3 in Japan ranged from 6,000 yen/m^3 to 19,000 yen/m^3 (from 46 Euro/m^3 to 146 Euro/m^3). He also mentioned that the OPEX of a landfill site is approximately 11,000 yen/m^3 (85 Euro/m^3). This indicates that the gate fee of a landfill site ranges from 17,000 yen/m^3 to 30,000 yen/m^3 (from 131 Euro/m^3 (118 Euro/t-residue) to 231 Euro/m^3 (208 Euro/t-residue)). ISWA Working Group on Thermal Treatment of Waste (2006) also reported landfill costs with tax, which varied widely depending on the country. According to the report, the gate fees in Italy, Norway and Sweden ranged from 85 to 110 Euro/t, from 105 to 140 Euro/t and from 75 to 130 Euro/t, respectively. On the other hand, it also reported that other countries such as Germany or France have much cheaper gate fees (approximately 50 Euro/t and 60 Euro/t). CEWEP country reports (2010) reported similar landfill gate fees. In this study, the gate fee of the landfill site is set as 118 Euro/t, which is the same as the lower limit of those reported by Higuchi (2006). The gate fee of fly ash is set to be 250 Euro/t. Slag and metal produced in the cogasification system assumed to have no disposal cost, though they were sold as valuable resources in Japan.

The power consumption of the landfilling and diesel consumption for the vehicles are set as 64.5 kWh/t-residue and 0.62 L/t-residue, respectively (Osada et al., 2012). The diesel oil price is set as 1.4 Euro/L (Europe's Energy Portal, 2013). The effect of transportation of residue between the sites was neglected in this study. The parameters of cases 4 to 6 are the same as those of cases 1 to 3.

TABLE 7: Parameter list

		Scheme A			Scheme B		
		Case 1	Case 2	Case 3	Case 4	Case 5	Case 6
	unit		Incineration			Incineration	
		DMS	Without ash melting	With ash melting	DMS	Without ash melting	With ash melting
Coke	M Euro /a	1.62	0.00	0.00	1.37	0.00	0.00
Natural Gas	M Euro /a	0.07	0.05	0.05	0.06	0.05	0.05
Power	M Euro /a	-4.23	-3.71	-0.86	-3.59	-3.71	-1.82
Other Utilities	M Euro /a	2.46	0.00	1.95	2.09	0.00	1.95
Bottom Ash Disposal	M Euro /a	0	5.44	1.67	0	3.97	1.47
Fly Ash Disposal	M Euro /a	1.35	0.98	0.98	1.15	0.98	0.98
Slag Recycling	M Euro /a	0	0	0	0	0	0
Metal Recycling	M Euro /a	0	0	0	0	0	0
Power for Landfilling	M Euro /a	0.02	0.19	0.07	0.02	0.15	0.06
Diesel for Vehicles	M Euro /a	0.00	0.04	0.02	0.00	0.03	0.01
Total	M Euro /a	1.29	3.00	3.88	1.10	1.47	2.70
Operating Cost*1	M Euro	26	60	78	22	29	54
Differences	M Euro	-	34	56	-	7	32

**1: Operating cost was calculated in relation to 20 years operation utilities.*

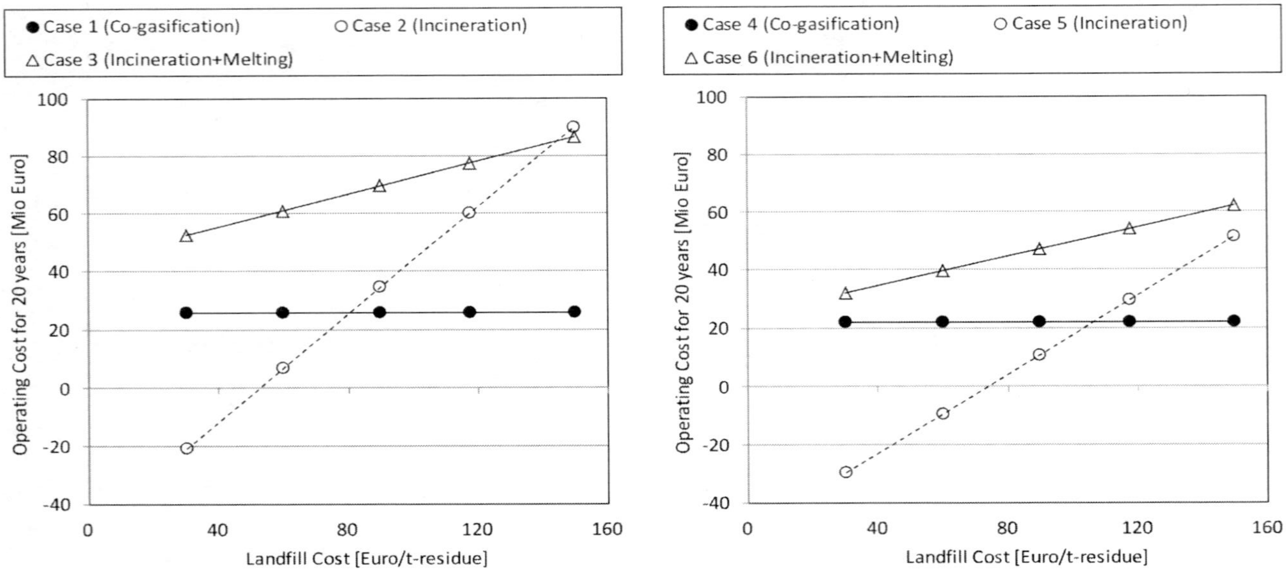

Scheme A

Waste Compositions: NCV 9.0 MJ/kg, Inert contents 11.8%
Co-gasification of MSW with bottom ash discharged from another plant
Gross Power Generation: 23%, Power Price: 60 Euro/t-waste
Power Consumption: DMS 170 kWh/t, Incineration 100 kWh/t,

Scheme B

Waste Compositions: NCV 9.0 MJ/kg, Inert contents 22.0%
Gross Power Generation: 23%, Power Price: 60 Euro/t-waste
Power Consumption: DMS 170 kWh/t, Incineration 100 kWh/t,

FIGURE 2: Effects of landfill cost on operating cost

5.3.2.3 COST EVALUATION OF THE CO-GASIFICATION SYSTEM

Table 7 shows the results of the cost calculations. In case 1, the costs of other utilities and coke were 2.46 million Euro/annum and 1.62 million Euro /annum. The income from power generation was 4.23 million Euro /annum. The cost of fly ash disposal was 1.35 million Euro /annum. The cost of bottom ash disposal (0 Euro/annum) was much lower than that in other cases. Other costs were negligible. In case 2, the largest annual cost was landfill cost (5.4 million Euro). The income from power selling was 3.7 million Euro. In the case 3, the income of power generation was 0.86 million Euro, which was much lower than in other cases. This is caused by the power consumption of the ash melting system. The cost of bottom ash disposal was between those in cases 1 and 2. Cases 4 to 6 have almost the same tendencies as those in cases 1 to 3.

The annual costs in cases 1, 2 and 3 were 1.29, 3.00 and 3.88 million Euro, respectively. In this study, operating cost was evaluated. This operating cost was calculated as the 20-year operating cost, including the above-mentioned items. The CAPEX, maintenance cost and labor fee were not included. Case 1 had the minimum operating cost among scheme A, at 26 million Euro. This figure was 60 million Euro and 78 million Euro cheaper than those in cases 2 and 3, respectively. In scheme B, the annual costs in cases 4, 5 and 6 were 1.10, 1.47 and 2.70 million Euro, respectively. Compared to pattern 1, the cost advantage of the co-gasification system was low. However, from the operating cost point of view, the co-gasification system was 7 and 32 million Euro cheaper than those in cases 5 and 6, respectively.

These results indicate that the co-gasification system with the DMS technology has the potential to reduce operation costs through minimizing the final landfill amount and maximizing material recovery from waste in these boundary conditions.

5.3.3 SENSITIVITY ANALYSESOF LANDFILL GATE FEE

The effect of landfill gate fee on operating cost for 20 years are shown in Figure 2. As mentioned above, the landfill gate fee is an important criterion in evaluating the waste management scheme. The landfill cost varies from 30 to 150 Euro/t-residue. Other parameters such as coke consumption, power price, or power generation efficiency are kept as the same. Waste compositions such as inert materials and NCV are also kept as the same.

By increasing the landfill gate fee of bottom ash, the operating cost increased in all the cases. In scheme A, the cheapest system varied depending on the landfill gate fee. When the gate fee was higher than 80 Euro/t-residue, the co-gasification system of case 1 was the cheapest system among the three cases. When the gate fee was higher than 150 Euro/t-residue, the operating cost of incineration with ash melting (case 3) was lower than that of case 2. In the cheaper range of landfill gate fee, incineration (case 2) showed the lowest operating cost among the three cases. Scheme B also had operating cost variation depending on the landfill gate fee. When the landfill gate fee was higher than 105 Euro/t-residue, the co-gasification system showed the lowest operating cost. Compared to incineration with an ash melting (case 6), the co-gasification system was cheaper in all ranges.

The effect of the landfill gate fee varied depending on the system. In both the schemes, incineration cases (cases 2 and 5) were highly influenced by the landfill gate fee. In this study, bottom ash recycling via the aging process is not allowed, due to toxic heavy metal concentrations. Bottom ash is directly transferred to landfill sites. This causes the strong influence of the landfill gate fee. On the other hand, the co-gasification system was not affected by the gate fee. This was caused by the final landfill amount. The co-gasification system with the DMS technology can minimize the landfill amount and maximize resource recovery from waste. Therefore, the landfill gate fee does not greatly affect the operating cost of the co-gasification system.

These results indicate that the co-gasification system with the DMS technology is suitable especially for areas with high landfill gate fees. When the slag and metal produced in the DMS are sold as valuable resources as well as in Japan, the operating cost of the co-gasification system could be improved. In addition to the price of resources recovery, the coke consumption of the co-gasification system is the important parameter. It is reported that the latest technology applied to the DMS can reduce the coke consumption to 25 kg/t-residue (Tanigaki et al., 2013 b). When the coke consumption is reduced from 44.6 kg/t-waste to 25 kg/t-waste, the cost of coke consumption is reduced from 1.62 million Euro/annum to 0.90 million Euro/annum. As a consequence, the co-gasification system has an operational advantage in the range that the gate fee of landfill is higher than 60 Euro/t-residue.

5.5 CONCLUSIONS

In this study, a new waste management scheme using the co-gasification system of MSW with bottom ash and incombustibles was evaluated, based on operating data and the operating cost for 20 years under certain boundary conditions.

From a technical point of view, the co-gasification system using the gasification and melting technology (the Direct Melting System) stably processed MSW with bottom ash and incombustibles from a recycling center together. In addition, the system produced slag and metal containing few toxic heavy metals such as lead and zinc. Because of the compositions of the slag, it can be recycled as secondary materials. These results indicate that molten materials produced are inert and the co-gasification system using the DMS technology can minimize final landfill amounts via recovering inert materials of waste processed as slag and metal.

The economic impact of the co-gasification system was evaluated for two waste management schemes in a certain boundary condition. In both the schemes, the co-gasification system was compared with alternatives such as incineration technologies with and without an ash melting system. The incineration with the ash melting system can convert the bottom ash into inert materials and reduce the final landfill amount as well as the co-gasification system. Scheme A processed MSW containing low ash contents, bottom ash from other incineration plants and incombustibles from a recycling center. In this scheme, the co-gasification system using the DMS showed an economic advantage over other systems. Scheme B processed incombustibles and MSW containing high ash contents. The co-gasification system also had an economic advantage over other systems.

Sensitivity analyses were also conducted to clarify the effects of landfill gate fee on operating cost. In the variation of the landfill cost, the co-gasification system had the advantage especially in the case of higher landfill gate fees, at 80 Euro/t-residue or more in scheme A and 105 Euro/t-residue or more in scheme B. Incineration with the ash melting system was the most expensive system in both schemes.

These results indicate that the co-gasification of bottom ash and incombustibles with MSW contributes to minimizing the final landfill amount and has great possibilities for maximizing material recovery and energy recovery from waste. Especially, this co-gasification system shows great advantage in cases of high landfill gate fees.

REFERENCES

1. Aigner I., Pfeifer C. and Hofbauer H. (2011). Co-gasification of coal and wood in a dual fluidized bed gasifier. Fuel, vol. 90, 2404-2412.
2. Arena U. and Di Gregorio F. (2013) (a). Gasification of a solid recovered fuel in a pilot scale fluidized bed reactor. Fuel, vol. 117, Part A, n. 30, 528-536.
3. Arena U. and Di Gregorio F. 2013 (b). Element partitioning in combustion- and gasification-based waste-to-energy units. Waste Manage., vol. 33, 1142-1150
4. Arena U., Zaccariello L. and Mastellone M.L. (2010). Fluidized bed gasification of waste-derived fuels. Waste Manage., vol.30, 1212–1219.
5. Birgisdottir H., Bhander G., Hauschild M.Z. and Christensen T.H. (2007). Lifecycle assessment of disposal of residues from municipal solid waste incineration: Recycling of bottom ash in road construction or landfilling in Denmark evaluated in the ROAD-RES model. Waste Manage., vol. 27, S75-S84.
6. Confederation of European Waste-to-Energy Plants (CEWEP) (2010). Country Report on Waste Management. http://www.cewep.eu/information/data/subdir/442._Country_Report_on_Waste_Management.html
7. De Boom A., Degrez M., Hubaux P. and Lucion C. (2011). MSWI boiler fly ash: Magnetic separation for material recovery Waste Manage., vol. 31, 1505-1513.
8. Europe's Energy Portal (2013). www.energy.eu/fuelprices/.
9. European Commission (2006). Integrated Pollution Prevention and Control, Reference Document on the Best Available Techniques for Waste Incineration. August, 2006.
10. European Union (1999). Council Directive 1999/31/EC of 26 April 1999 on the landfill of waste. Official Journal of the European Communities.
11. Eurostat (2013). Gas prices for industrial consumers. http://epp.eurostat.ec.europa.eu/tgm/table.do?tab=table&init=1&language=en&pcode=ten00112&plugin=0
12. Higuchi, S. Ishida Y. and Hanashima M. (2006). Research on Resource Stock Type Landfill System (in Japanese). Proceedings of 17th Annual Congress of Japan Society of Material Cycles and Waste Management, 598-600.

13. https://www.rite.or.jp/Japanese/labo/sysken/about-global-warming/download-data/EIA_waste_management.pdf
14. Hyks J., Astrup T. and Christensen T.H. (2009). Leaching from MSWI bottom ash: Evaluation of non-equilibrium in colum percolation experiments. Waste Manage., vol. 29, 522-529.
15. International Solid Waste Association (ISWA) Working Group on Thermal Treatment of Waste, 2006. "Management of Bottom Ash from WTE Plants", An overview of management options and treatment methods. http://www.iswa.org/uploads/tx_iswaknowledgebase/Bottom_ash_from_WTE_2006_01.pdf.
16. Manako K., Kashiwabara T., Kobata H., Osada M., Takeuti S. and Mishima T. (2007). Dioxins Control and High-Efficiency Power Gereation in a Large-Scale Gasification and Melting Facility. Proceedings of DIOXIN 2007 International Symposium, 940-943
17. Mastellone M.L. and Arena U. (2008). Olivine as a tar removal catalyst during fluidized bed gasification of plastic waste. AIChE Journal, vol. 54, n. 6, 1656-1667.
18. Mastellone M.L., Zaccariello L. and Arena U. (2010). Co-gasification of coal, plastic waste and wood in a bubbling fluidized bed reactor. Fuel, vol. 89, 2991-3000.
19. Oda T and Satoh T. (2010). Suita Municipal Waste Incineration Facility (in Japanese). Takuma Technical Review, vol. 18, n. 1, 19-24.
20. Osada M., Manako K., Hirai Y., and Sakai S. (2012). Life Cycle Assessment for Treatment and Recycling of Automobile Shredder Residue (ASR) (in Japanese). Journal of Japan Society of Material Cycles and Waste Management, vol. 23, n. 6, 264-278.
21. Osada S., Kuchar D. and Matsuda H. (2009). Effect of chlorine on volatilization of Na, K, Pb, and Zn compounds from municipal solid waste during gasification and melting in a shaft-type furnace. J Mater Cycles Waste, vol. 11, 367-375.
22. Osada S., Kuchar D. and Matsuda H. (2010). Thermodynamic and experimental studies on condensation behavior of low boiling-point elements volatilized in the melting process. J Mater Cycles Waste, vol. 12, 83–92.
23. Osada, M. Tanigaki, N., Takahashi, S., Sakai. S., 2008. Brominated flame retardants and heavy metals in automobile shredder residue (ASR) and their behavior in the melting. Journal of Material Cycles and Waste Management, 93-101
24. Pinto F., Franco C, Andre R.N., Miranda I, Gulyurtlu I and Cabrita I (2008). Co-gasification study of biomass mixed with plastic wastes. Fuel, vol. 81, n. 3, 291–297.
25. Research Institute of Innovative Technology for the Earth, 2010. Comprehensive Assessment of the Environmental Impacts and the Cost in Waste Management System (in Japanese).
26. Tanigaki N., Fujinaga Y., Kajiyama H., Ishida Y. (2013a). Operating and environmental performances of commercial-scale waste gasification and melting technology. Waste Manage. Res., vol. 31, n. 11, 1118-1124.
27. Tanigaki N., Manako K. and Osada M. (2012). Co-gasification of Municipal Solid Waste and Material Recovery in a Large-scale Gasification and Melting System. Waste Manage., vol. 32, 667–675.
28. Tanigaki, N., Yoshimoto, Y., Ishida, Y., Osada, M., 2013b. Effects of Air Pre-heating for Combustible Dust Injection on Municipal Solid Waste Gasification and Melting

System. Proceedings of 14th International Waste Management and Landfill Symposium (Sardinia 2013), Sep. 30 – Oct. 4, 2013, S. Margherita di Pula, Italy

29. Taylor R., Ray R. And Chapman C. (2013). Advanced thermal treatment of auto shredder residue and refuse derived fuel. Fuel, vol. 106, 401-409.
30. Willis K. P., Osada S. and Willerton K. L. (2010). Plasma Gasification: Lessons learned Ecovalley WTE Facility. Proceedings of the 18th Annual North American Waste-to-Energy Conference (NAWTEC 18), May 11-13, 2010, Orlando, Florida, USA.

CHAPTER 6

Analysis of Organic and Inorganic Contaminants in Dried Sewage Sludge and By-Products of Dried Sewage Sludge Gasification

SEBASTIAN WERLE AND MARIUSZ DUDZIAK

6.1 INTRODUCTION

Gasification is regarded as a prospective and promising method for rendering sewage sludge (an example of unconventional biomass) harmless [1–4]. Unfortunately, apart from the valuable gas fuel produced, the process results in solid and liquid waste by-products [5–7]. The solidification of mineral substances during gasification produces solid products, mostly ash, but also char coal in some cases [5,6]. The formation of the latter depends on the composition of nonflammable inorganic substances in the sludge which considerably decreases the Debye characteristic temperatures of the ash [6]. Liquid products, i.e., tar, as a result of the condensation of the contaminants present in gas [7].

Sewage sludge, apart from energetically desirable compounds, is the source of toxic and hazardous organic and inorganic contaminants. Organic compounds identified in it include [8–11]: dioxins and furans, polychlorinated biphenyls (PCBs), organochlorine pesticides, adsorbed and extracted chloro derivatives, polycyclic aromatic hydrocarbons (PAHs), phenols and their derivatives, phthalates, sex hormones and others. The group of hazardous inorganic compounds assayed in sewage sludge contain primarily various heavy metals at a wide range of concentrations (mg/kg dry basis for raw sewage) [12–18]: arsenic 3–230, cadmium 1–3410, chromium 10–990,000, copper 80–2300, nickel 2–179, lead 13–465, and zinc 101–49,000.

As mentioned before, dried sewage sludge gasification produces waste by-products. This phenomenon accompanies all thermal treatment processes of both traditional and unconventional biomass (incineration, co-incineration and others) [1,3]. Thermal techniques usually transfer and accumulate contaminants in liquid and solid phases, which do not eliminate environmental hazards. It may be especially harmful while thermally processing sewage waste which is originally a source of various toxic and hazardous compounds. The preliminary research [19] found that the toxic effect of products produced during sewage sludge gasification depends on both the type of a sample tested (ash, char coal and tar) and sewage sludge used. That research was carried out using a Microtox® test with a *Vibrio fischeri* bacterial strain (luminescence method). Higher toxicity was found in the samples of ash that formed during gasification of sewage sludge, which appeared to be toxic, than for sludge of lower toxicity. As for tar samples, they were all toxic, regardless of the sludge gasified. Thus, from a cognitive perspective, the characteristics of such samples in terms of different contaminants which might be responsible for their toxicity are significant.

Taking into account the abovementioned aspects, this research covered multidirectional chemical instrumental activation analyses of dried sewage sludge and waste by-products formed during its gasification (ash, char coal and tar) in a fixed bed reactor. The tests were aimed at assessing transport and transformations of organic and inorganic contaminants during sewage sludge gasification in the sludge-gasification-solid and liquid waste by-products system.

FIGURE 1: Sewage sludge analyzed: (a) Sewage Sludge 1—from mechanical and biological wastewater treatment plants (WWTP); (b) Sewage Sludge 2—mechanical, biological and chemical WWTP with simultaneous phosphorus precipitation.

6.2 MATERIALS AND METHODS

6.2.1 MATERIALS

Two different sewage sludges taken from wastewater treatment plants (WWTP) located in the West and North of Poland were selected for the research. Sewage Sludge 1 came from a WWTP operating in a mechanical and biological system while Sewage Sludge 2 was collected in a mechanical, biological and chemical WWTP with simultaneous phosphorus precipitation. The sludge produced at the plants was subject to fermentation and then, after dewatering, dried in a cylindrical drier on shelves heated up to 260 °C (Sewage Sludge 1) and using hot air at a temperature of 150 °C in a belt drier (Sewage Sludge 2). As a result, Sludge 1 took the form of granules while Sludge 2 was in the shape of irregular thin "pasta" (Figure 1).

The sludge was gasified in a fixed bed reactor using air as a gasifying agent fed to the reactor at a temperature of 25 °C and the amount of the agent corresponded to the ratio of excess air (λ) 0.18. The effect of the gasification parameters on producer gas yield, its composition and calorific value in particular, are discussed in detail in [6,20,21]. As far as the by-products are concerned, the gasification of Sewage Sludge 1 produced both ash (taken from the ash-pan) and char coal (taken from the inside of the reactor), while Sewage Sludge 2 produced ash only. The char coal formation in the case of the Sewage Sludge 1 was caused by a noncombustible inorganic compound. Those contaminants cause a significant reduction of the ash fusion temperatures. This influence is especially visible in the value of the temperature of the initial deformation [7].

6.2.2 ANALYSIS OF SELECTED QUALITY INDICATORS OF SEWAGE SLUDGE AND GASIFICATION BY-PRODUCTS

The chemical qualitative analysis of the dried sewage sludge and waste by-products produced during gasification covered the concentrations of major elements (carbon, hydrogen, nitrogen, chlorine, fluorine, sulphur and oxygen), mineral elements (magnesium, calcium), phosphorus, alkaline metals (sodium and potassium), and selected heavy metals (zinc, selenium, lead, nickel, mercury, copper, chromium, cadmium and arsenic). The concentrations of the major elements were assayed, using automatic measurements with an infrared (IR) analyzer. Both the mineral elements and heavy metals were determined by plasma or absorption spectrometry. The assays carried out in the liquid products covered: total organic carbon (TOC) using an automatic analyzer, conductivity (conductometric analysis) and ammonia nitrogen concentration (spectrophotometry).

TABLE 1: Sewage sludge properties.

Element/Parameter		Sewage Sludge 1	Sewage Sludge 2
Proximate analysis *, % (as received)	Moisture	5.30	5.30
	Volatile matter	51.00	49.00
	Ash	36.50	44.20
Ultimate analysis, % (dry basis)	C	31.79	27.72
	H	4.36	3.81
	N	4.88	3.59
	O (by difference)	20.57	18.84
	S	1.67	1.81
	F	0.013	0.003
	Cl	0.22	0.03
Calorific value *	Higher heating value (HHV), MJ/kg dry basis	14.05	11.71
	Lower heating value (LHV), MJ/kg dry basis	12.96	10.75

**Based on the works [6,20,21]*

The authors of earlier works in this field [6,20,21] assayed moisture, volatile fractions, ash, heat of combustion and their calorific value as well. As for the ash, indicators characteristic of its tendency towards slagging and fouling of heated surfaces as well as agglomerate formation were determined [22]. The assays of moisture, volatile fractions and ash were conducted with the gravimetric methods given in the following standards PN-EN 14774-3:2010 [23], PN-EN 15402:2011 [24] and PN-EN 15403:2011 [25], respectively. The heat of combustion was determined by the calorimetric method while the calorific value was calculated, using the mass fractions of the major elements in a sample. The characteristics of two sewage sludges are given in Table 1.

6.2.3 DETERMINATION OF LOW-MOLECULAR ORGANIC COMPOUNDS WITH GAS CHROMATOGRAPHY-MASS SPECTROMETRY (GC-MS) TECHNIQUE

The sewage sludge samples, as well as solid and liquid gasification waste by-products, were qualitatively and quantitatively analyzed using GC-MS to assess their contamination with low-molecular organic compounds. The assays were targeted at four main groups of contaminants commonly identified in sewage sludge [9,11], i.e., PAHs, pesticides, PCBs, phenols and their derivatives. The analysis employed the following Sigma-Aldrich (Poznan, Poland) standard solutions:

- PAHs solution which contained 16 compounds [acenaphthene, acenaphthylene, anthracene, benzo(a)anthracene, benzo(a)pyrene, benzo(a)fluoranthene, benzo(b)fluoranthene, benzo(g,h,i)perylene, benzo(k)fluoranthene, chrysene, dibenzo(a,h)anthracene, phenanthrene, fluorene, indeno(1,2,3-cd)pyrene), naphthalene, pyrene] in a concentration of 100 ng/μL in toluene;
- A solution of sixteen pesticides (aldrin, α-BHC, β-BHC, σ-BHC, dieldrin, α-endosulfan, β-endosulfan, endosulfan sulfate, endrin, endrin aldehyde, γ-BHC, heptachlor, heptachlor epoxide, 4,4′-DDE, 4,4′-DDE, 4,4′DDT) in a concentration of 200 ng/μL in n-hexane;
- A solution of PCBs which contained six different derivatives [No. 28 (2,4,4′-PCB), 52 (2,2′,5,5′-PCB), No. 101 (2,2′,4,5,5′-PCB), No. 138 (2,2′,3,4,4′,5-PCB), No. 153 (2,2′,4,4′,5,5′-PCB), No. 180 (2,2′,3,4,4′,5,5′-PCB)] in a concentration of 10 ng/μL of the particular compounds prepared in isooctane;
- A solution of ten phenols and their derivatives (pentachlorophenol, 2-methyl-4,6-dinitrophenol, 2-chlorophenol, 2-nitrophenol, 2,4-dichlorophenol, 2,4-dimethylphenol, 2,4-dinitrophenyl, 2,4,6-trichlorophenol, 4-chloro-

3-methylphenol, 4-nitrophenol) in a concentration from 100 ng/μL to 250 ng/μL in methanol.

As a preliminary step in the chromatography assays, the solid samples were extracted with an organic solvent supported by ultrasounds. The analytical samples of sludge, ash or char coal (from 100 mg to 200 mg) were mixed with methylene chloride (1 mL) and placed in an ultrasonic bath (30 min). The extract produced was thickened and analyzed by GC-MS. The tar samples were thinned down to 10% in n-hexane and then analyzed chromatographically.

The chromatography analyses were performed using: a gas chromatograph coupled with a mass spectrometer (Saturn 2100 T GC-MS Varian, Warsaw, Poland) and equipped with a Supelco SLB™-5 ms column (30 m × 0.25 mm internal diameter, 0.25 μm film thickness), a split injector which maintained a constant temperature of 240 °C, helium (5 N) as a carrier gas with a flow rate of 1.1 mL/min. The temperature settings of the chromatography oven were as follows: 50 °C (4 min) → 8 °C /min → 260 °C → 4 °C/min → 300 °C (5 min). The temperature of the ion trap and ion source was 200 °C. The ions were within the m/z range of 40 to 400.

6.3 RESULTS AND DISCUSSIONS

6.3.1 GENERAL CHARACTERISTICS OF GASIFIED SEWAGE SLUDGE

A comparison of the sewage sludge analyzed pointed to the conclusion that Sewage Sludge 1 exhibited a higher calorific value than Sewage Sludge 2 (Table 1). This was also confirmed by the analysis of gas composition and its calorific value shown in our papers [6,20,21]. However, the calorific value of the sewage sludge is lower than that of traditional biomass (wheat straw, rape straw, the sawdust of common osier, pine and oak) [6]. Our other paper [22] also compared the chemical composition of ash, sewage sludge and traditional biomass. It has been found that the sewage sludge exhibited much higher concentration of iron, titanium, phosphorus compounds and lower concentrations of potassium compounds. The above characteristics affect the typical indicators of ash depicting its tendency to slag and foul heated surfaces as well as form agglomerates. Compared with traditional biomass, sewage sludge tends to foul surfaces used for thermal treatment much less, but unfortunately, tends to slag and form agglomerates much more.

6.3.2 CHARACTERISTICS OF SEWAGE SLUDGE AND ITS GASIFICATION WASTE BY-PRODUCTS IN TERMS OF LOW-MOLECULAR ORGANIC COMPOUNDS

The chromatographic analyses of the dried sewage sludge extracts conducted herein confirmed their contamination with low-molecular organic compounds, PAHs in particular (Figure 2). Table 2 shows the concentrations of compounds assayed in the sewage sludge.

As mentioned above, the groups of organic contaminants assayed in the sewage sludge contained primarily PAHs. Sewage Sludge 1 revealed nine [phenanthrene, anthracene, benzo(a)fluoranthene, pyrene, chrysene, benzo(b)fluoranthene, dibenzo(a,h)anthracene, benzo(g,h,i)perylene, indeno(1,2,3-cd)pyrene], while Sewage Sludge 2 revealed eight [acenaphthene, benzo(a)fluoranthene, pyrene, benzo(a)anthracene, chrysene, benzo(b)fluoranthene, benzo(a)pyrene, indeno(1,2,3-cd)pyrene] compounds from that group of contaminants [naphthalene, acenaphthylene, acenaphthene, fluorene, phenanthrene, anthracene, benzo(a)fluoranthene, pyrene, benzo(a)anthracene, chrysene, benzo(b)fluoranthene, benzo(k)fluoranthene, benzo(a)pyrene, dibenzo(a,h)anthracene, benzo(g,h,i)perylene, indeno(1,2,3-cd)pyrene]. However, the total concentration of PAHs (Table 2) was four times higher in Sewage Sludge 1 than in Sewage Sludge 2. The papers [8,9] also found that PAHs constitute the basic group of contaminants in sewage sludge.

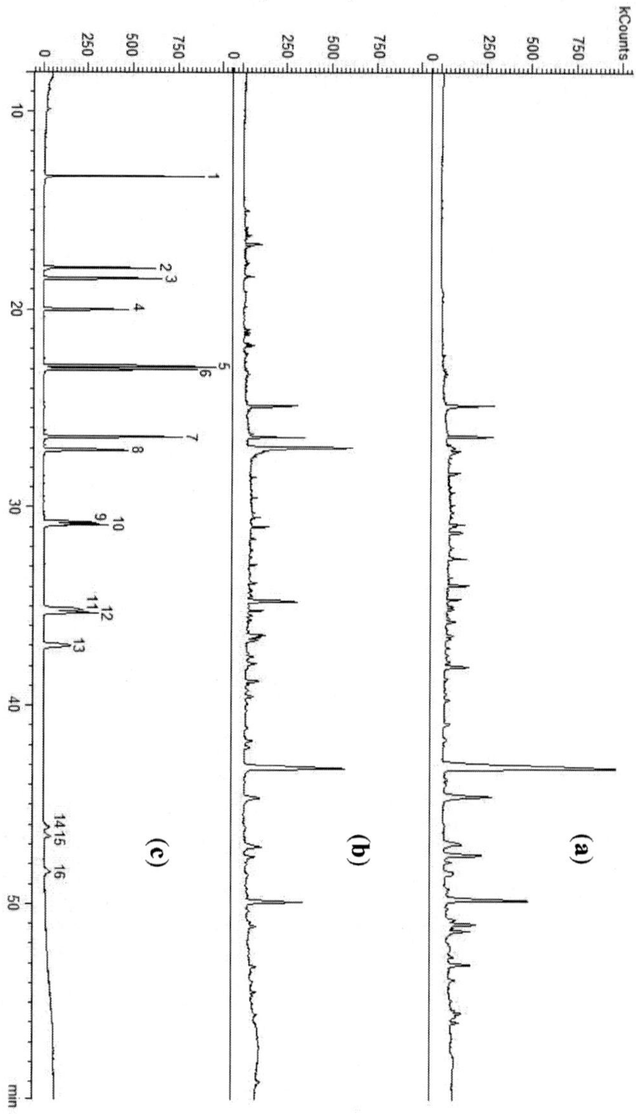

FIGURE 2: Chromatograms plotted after the analysis of dried sewage sludge extracts: (a) Sewage Sludge 1; (b) Sewage Sludge 2; and (c) 16 PAHs standard solution: (1) naphthalene; (2) acenaphthylene; (3) acenaphthene; (4) fluorene; (5) phenanthrene; (6) anthracene; (7) benzo(a)fluoranthene; (8) pyrene; (9) benzo(a)anthracene; (10) chrysene; (11) benzo(b)fluoranthene; (12) benzo(k)fluoranthene; (13) benzo(a)pyrene; (14) dibenzo(a,h)anthracene; (15) benzo(g,h,i)perylene; and (16) indeno(1,2,3-cd)pyrene.

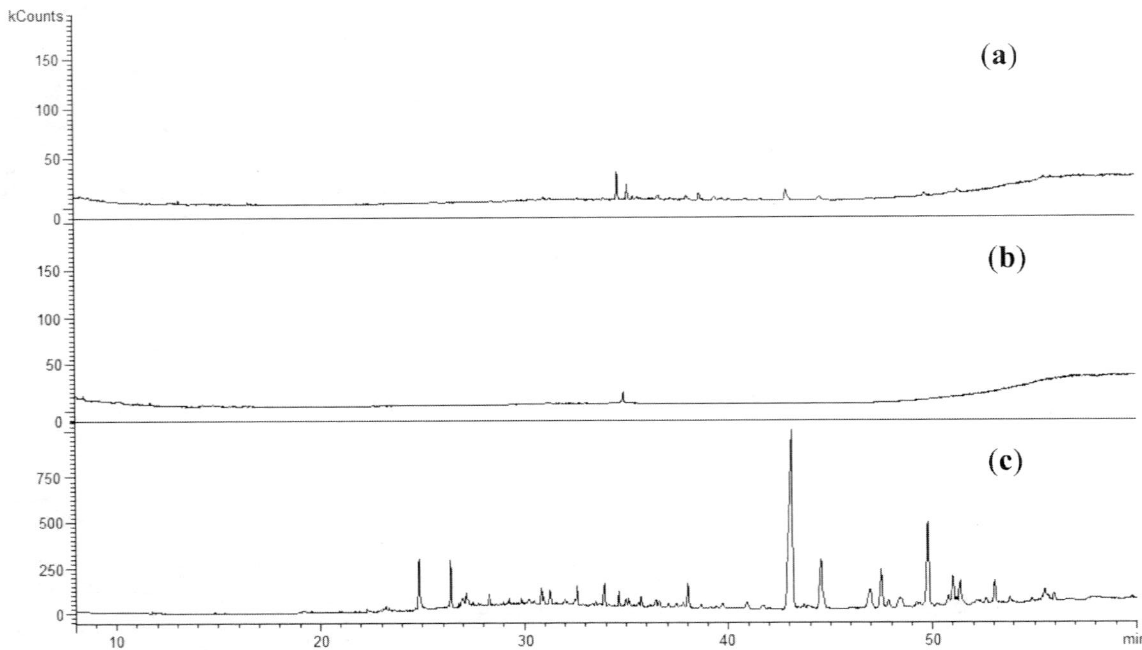

FIGURE 3: Chromatograms plotted after the analysis of (a) ash; (b) char coal; and (c) dried gasified Sewage Sludge 1 extracts.

TABLE 2: Concentration of organic compounds in sewage sludge. PAHs: polycyclic aromatic hydrocarbons; PCBs: polychlorinated biphenyls.

Compound		Retention time	Concentration, μg/kg dry basis	
			Sewage Sludge 1	Sewage Sludge 2
PAHs	acenaphthene	18.45	n.d.	80.84
	phenanthrene	22.89	511.12	n.d.
	anthracene	23.03	200.03	n.d.
	benzo(a)fluoranthene	26.46	44.78	126.48
	pyrene	27.08	187.22	123.86
	benzo(a)anthracene	27.13	n.d.	35.15
	chrysene	30.76	108.14	23.79
	benzo(b)fluoranthene	30.90	700.51	53.62
	benzo(a)pyrene	37.05	n.d.	46.11
	dibenzo(a,h)anthracene	46.13	101.54	n.d.
	benzo(g,h,i)perylene	46.61	209.44	n.d.
	indeno(1,2,3-cd)pyrene	48.39	370.62	131.48
	Sum	-	2,433.40	621.33
Pesticides	heptachlor	24.16	4.14	n.d.
	aldrin	25.04	3.13	1.28
	endrin	28.09	11.58	n.d.
	Sum	-	18.85	1.28
PCBs	2,2′,5,5′-PCB	24.79	9.75	7.90
	2,2′,4,5,5′-PCB	28.82	33.33	n.d.
	2,2′,4,4′,5,5′-PCB	29.47	23.78	4.57
	Sum	-	66.86	12.47

n.d.: Not detected

The group of organic contaminants identified in the sewage sludge also contained pesticides and PCBs. It is worth mentioning that Sewage Sludge 1 revealed the presence of three pesticides (heptachlor, aldrin and endrin) from the group of those commonly found in the natural environment in Poland (heptachlor, hexachlorocyclohexane, heptachlor epoxide, aldrin, endrin) [26]. Like the group of PAHs, the total concentrations of pesticides and PCBs were higher in Sewage Sludge 1 than in Sewage

Sludge 2. The sewage sludge samples did not reveal any low-molecular phenols or their derivatives.

The profile of the organic contaminant concentrations obtained for the sewage sludge was typical of the sludge produced during treatment of domestic wastewater with an addition of industrial wastewater [11]. Obviously, considering the concentrations of particular contaminant groups, the contribution of industrial wastewater was higher for Sewage Sludge 1 than for Sewage Sludge 2. This is also supported by the heavy metal concentrations, which will be discussed in the following.

Figure 3 gives chromatograms prepared after analyzing the extracts of solid waste by-products, i.e., ash, char coal and gasified dried sewage sludge 1. The plots revealed that the waste by-products were not contaminated with organic contaminants initially identified in the sewage sludge. The chromatograms of ash and char coal showed just a couple of peaks deriving from unidentified organic compounds. However, it was possible to notice that there were more peaks in the chromatograms for ash than char coal. This resulted from the fact that the ash also contained an organic fraction which is difficult to decompose thermally [27]. A similar correlation was found while performing a comparative analysis of an extract from Sewage Sludge 2 and ash produced during its gasification. The ash extract did not reveal any compounds present in the sewage sludge before the gasification.

The tests of the liquid products (tar) formed during sewage sludge gasification determined TOC which directly measures the amount of different organic substances in the sample. The indicator was very high for both the tar produced during gasification of Sewage Sludges 1 and 2, being 20,950 mg TOC/L and 22,390 mg TOC/L, respectively. A chromatographic analysis of the tar samples showed their contamination mainly by phenols and their derivatives (Table 3).

The formation of those contaminants during the gasification of coal, biomass and waste is a common phenomenon, although they are usually accompanied by other aromatic and polyaromatic compounds [10]. It was also observed that the total concentration of phenols and their derivatives was almost eight times higher in the tar produced during the gasification of Sewage Sludge 1 than the tar from Sewage Sludge 2. Since Sewage Sludge 1 initially contained more organic contaminants (Table 2), a correlation between the properties of the tar and gasified sewage sludge might be found.

TABLE 3: Concentrations of phenols and their derivatives in the tar produced during the gasification of the sewage sludge.

Compound	Retention time	Concentration, µg/L	
		Sewage Sludge 1	Sewage Sludge 2
2-Chlorophenol	9.28	211.84	57.20
2-Nitrophenol	12.55	89.52	n.d.
2,4-Dichlorophenol	12.95	361.56	n.d.
4-Chloro-3-Methylphenol	15.35	9.27	1.98
2,4,6-Trichlorophenol	16.33	62.32	46.33
Pentachlorophenol	22.50	57.97	53.90
Sum	-	792.48	100.23

n.d.: Not detected.

The drying temperature may also affect on the observed dependences. The drying temperature was much higher in the case of the Sewage Sludge 1 in comparison to Sewage Sludge 2. The higher drying temperature may affect on the tendency of the organic fraction from sewage sludge into the volatile organic compounds. As a result of the high drying temperature used sewage sludge has the properties similar to coal: firstly, hydrophobic properties by which storage of sludge is safer and without risk of the biological degradation and secondly, significantly improved regrind properties.

Phenols and their derivatives (Figure 4) are also identified in liquid waste by-products produced during biomass and waste gasification with water vapor [28–30]. It should also be mentioned that those components might act as a precursor of PAHs formation [31,32].

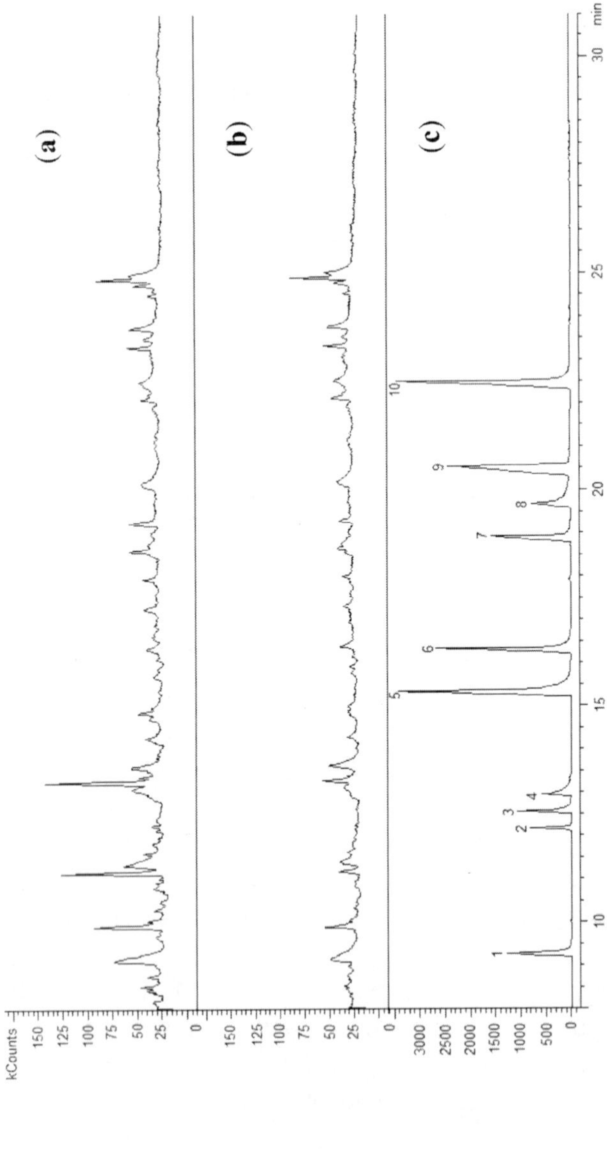

FIGURE 4: Chromatograms plotted after the analysis of the tar produced during sewage sludge gasification: (a) Sewage Sludge 1; (b) Sewage Sludge 2; and (c) the standard solution phenols and their derivatives: (1) 2-chlorophenol; (2) 2-nitrophenol; (3) 2,4-dimethylphenol; (4) 2,4-dichlorophenol; (5) 4-chloro-3-methylphenol; (6) 2,4,6-trichlorophenyl; (7) 2,4-dinitrophenol; (8) 4-nitrophenol; (9) 2-methyl-4,6-dinitrophenol; and (10) pentachlorophenol.(a)

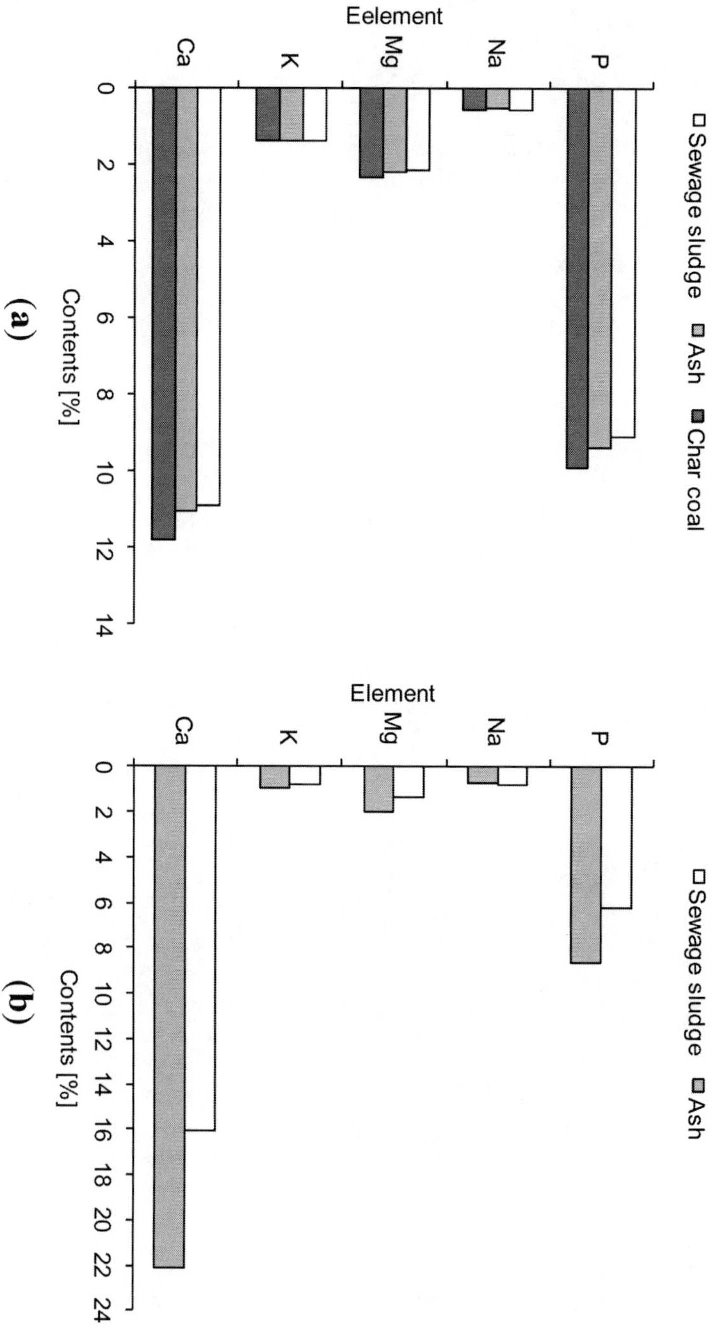

FIGURE 5: Mass fractions of selected mineral elements in the sludge: (a) Sludge 1; (b) Sludge 2 and in the solid products of gasification.

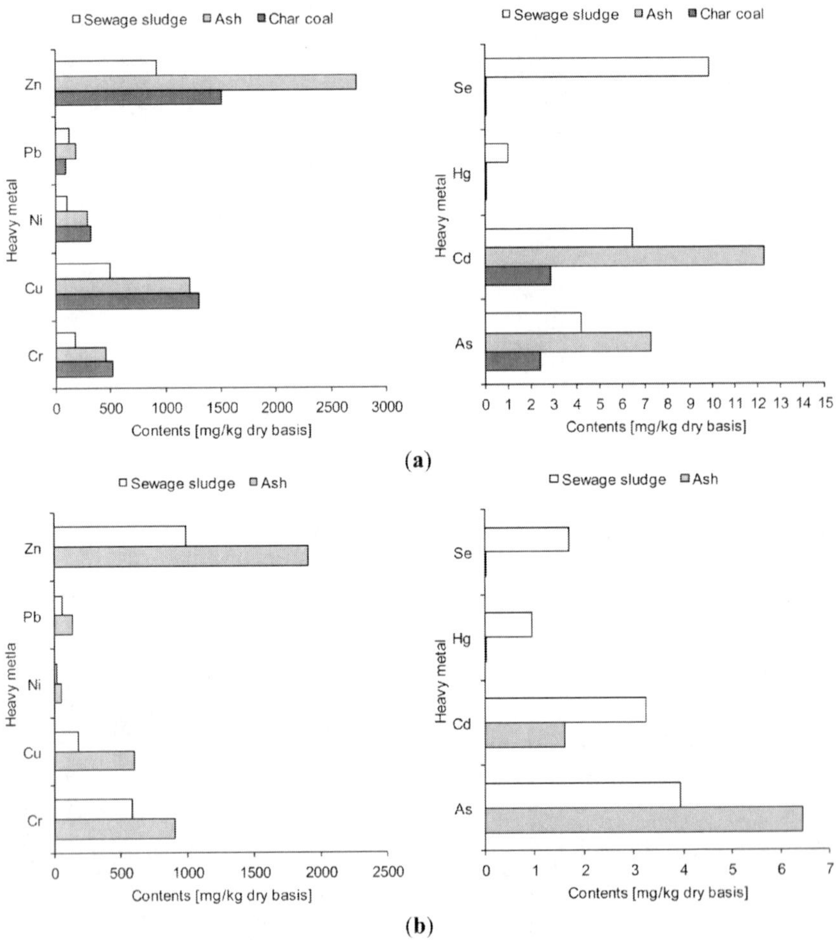

FIGURE 6: Concentrations of selected heavy metals in the sludge: (a) Sludge 1; (b) Sludge 2 and in the solid products of gasification.

6.3.3 ASSESSMENT OF SELECTED INORGANIC COMPOUNDS CONCENTRATIONS IN SEWAGE SLUDGE AND GASIFICATION WASTE BY-PRODUCTS

Thereafter, the paper assesses the concentrations of selected inorganic compounds in sewage sludge and gasification by-products.

As for the original contamination of the sewage sludge, it was different in terms of mass contribution of particular elements, including heavy metals (Figures 5 and 6). On the other hand, the total concentration of, e.g., heavy metals were similar in both sewage sludges, being 1841 mg/kg dry wt. for Sewage Sludge 1 and 1848 mg/kg dry wt. for Sewage Sludge 2.

It has been found that the mass contribution of phosphorus and calcium in the solid products was higher than that assayed in the sewage sludge prior to thermal treatment, the highest accumulation being observed for Sewage Sludge 2 and the ash produced during its gasification (Figure 5b). Similar effects are given in paper [33] which points out that phosphorus concentration in the ash formed after sewage sludge gasification increased from 14.1 mg/kg dry basis to 20.6 mg/kg dry basis (an increase by 68%).

Similarly, the concentrations of heavy metals were higher in ash and char coal than sewage sludge (Figure 6). This applied to seven (zinc, lead, nickel, copper, chromium, cadmium, arsenic) out of nine heavy metals investigated (zinc, lead, nickel, copper, chromium, selenium, mercury, cadmium and arsenic), irrespective of the type of sewage sludge. Similar conclusions were drawn by the authors of the paper [13] who found an increase in the concentrations of cadmium (from 0.93 to 1.67 mg/kg dry basis), chromium (from 80.82 to 247.95 mg/kg dry basis), copper (from 580.36 to 922.14 mg/kg dry basis), lead (from 78.27 to 125.09 mg/kg dry basis) and zinc (from 402.09 to 637.50 mg/kg dry basis) in sewage sludge and ash samples after gasification. However, some differences in the accumulation of the solid waste by-products of particular heavy metals were spotted. For instance, zinc concentration in an ash sample produced during Sewage Sludge 2 gasification increased twofold, and even three-fold for Sewage Sludge 1.

The heavy metals presence in solid by-products generated during sewage sludge gasification has an impact on their potential treatment or utilization. As it is well known, one of the main obstacles to agricultural usage of the sewage sludge is the high content of heavy metals, which can penetrate from the sewage sludge to the soil causing their contamination [16]. From the results of this study, it can be concluded that concentration of the heavy metals in solid by products from sewage sludge gasification process is higher in comparison to sewage sludge. It should be noted that there is an ecological risk from those type of waste. This was confirmed by the toxicity analysis presented in the introduction part of this work. Additionally, some possibilities of the solid gasification by products treatment are indicated in the Subsection 3.4.

TABLE 4: Comparison of the physical and chemical characteristics of the tar produced during sewage sludge gasification with current Polish permissible standards for contaminants in the sewage resulting from the thermal treatment of waste.

Indicator/heavy metal	Unit	Tar from sewage sludge gasification		The maximum concentration indicator for sewage sludge from the waste thermal treatment process [34]
		1	2	
pH	-	4.39	4.24	6.5–8.5
As	mg/L	0.29	0.16	0.15
Cd	mg/L	0.13	0.06	0.05
Cr	mg/L	0.02	0.01	0.50
Cu	mg/L	0.02	<0.01	0.50
Hg	mg/L	<0.01	<0.01	0.03
Ni	mg/L	0.01	<0.01	0.50
Pb	mg/L	0.38	0.17	0.20
Zn	mg/L	5.60	5.73	1.50

The liquid by-products (tar) produced during sewage sludge gasification demonstrated high conductivity, which proves their high contamination with various inorganic substances. The conductivity of the tar from Sewage Sludge 1 was 9800 µS/cm and 8170 µS/cm for Sewage Sludge 2. The hazardous inorganic contaminants occurring in the tar also included ammonia whose concentrations were 1090 mg NH_4^+/L (tar from Sewage Sludge 1) and 950 mg NH_4^+/L (tar from Sewage Sludge 2). Table 4 compares the physical and chemical characteristics of the tar produced during sewage sludge gasification with current Polish permissible standards for contaminants in the sewage resulting from the thermal treatment of waste in terms of, e.g., heavy metals [34]. As an effect of the condensation process in the end part of the gas pipe in the reactor, heavy metals from the gasification gas are transformed into tar.

It has been found that, out of eight heavy metals included in the government regulations (arsenic, cadmium, chromium, copper, mercury, nickel, lead and zinc), the concentrations of four metals (arsenic, cadmium, lead, zinc) in the tar from Sewage Sludge 1 and three metals (arsenic, cadmium and zinc) in the tar from Sewage Sludge 2 exceeded the standards. The same applies to the pH of the tar which fell outside the permissible range too (6.5–8.5).

Analyzing the sewage sludge and by-products after their gasification in terms of the heavy metals presence it is expected that gasification gas will be much less contaminated by these substances because the most of the inorganic substances originally present in the gasified sewage sludge undergoing accumulates in the solid by-products generated during the process. Such behaviors of heavy metals have been confirmed by other authors [13]. The organic compounds are transformed in the route of the complex thermochemical reactions accompanied by the completed decomposition of the compounds originally present in the sludge.

6.3.4 SUGGESTIONS FOR FURTHER RESEARCH

The authors suggest that further research should concentrate on the recovery of phosphorus from solid products (ash and char coal) formed during the gasification of sewage sludge, e.g., through leaching phosphorus from ash or sinter with mineral acids. The application of ash resulting from gasification to the sorption of toxic and hazardous compounds (e.g., heavy metals) from different sewage is also worth considering. Similar research has already been carried out, using dried sewage sludge [35,36]. As for the tar produced during sewage sludge gasification, it requires an individual treatment system since it is contaminated by phenols, their derivatives and heavy metals (causing considerable toxicity of the samples, which was described in the preliminary research in paper [19]).

6.4 CONCLUSIONS

The waste by-products (solid and liquid) produced during sewage sludge gasification differ in the type of contaminants present in them. The concentrations of the contaminants depended on the characteristics of the sewage sludge treated thermally. The organic compounds present in the initial sewage sludge were not identified in the solid products (ash and char coal). They were contaminated mainly with inorganic compounds, including heavy metals. The liquid gasification products (tar) contained both toxic and hazardous organic (phenols and their derivatives) and inorganic (heavy metals) compounds. Higher concentrations of contaminants were found in the by-products formed after the thermal treatment of sewage sludge which was initially considerably contaminated with different groups of organic (PAHs, pesticides and PCBs) and inorganic (heavy metals among others) compounds, although paradoxically, the sludge displayed a higher calorific value.

Gasification promotes the migration of certain inorganic compounds, e.g., phosphorus or zinc from the sewage sludge to the solid phase formed after the sludge was treated thermally. This depended on both the type of the solid by-product (ash and char coal) and the sewage sludge used. Taking all that into account, the valorization of solid gasification products to recover valuable compounds or use them in the sorption of toxic and hazardous sewage might be worth considering in the future.

The high accumulation of heavy metals, including chromium, in the solid by-products (ash and char coal) and, e.g., phenols and their derivatives in the liquid products (tar) points to the ecological hazard they create. In the case of the gasification tar, it is necessary to develop a separate treatment system.

The paper has been prepared within the frame of the National Science Centre project based on decision no DEC-2011/03/D/ST8/04035.

REFERENCES

1. Chun, Y.N.; Ji, D.W.; Yoshikawa, K. Pyrolysis and gasification characterization of sewage sludge for high quality gas and char production. J. Mech. Sci. Technol. 2013, 27, 263–272.
2. Kang, S.W.; Dong, J.I.; Kim, J.M.; Lee, W.C.; Hwang, W.G. Gasification and its emission characteristics for dried sewage sludge utilizing a fluidized bed gasifier. J. Mater. Cycles Waste Manag. 2011, 13, 180–185.
3. Werle, S.; Wilk, R.K. A review of methods for the thermal utilization of sewage sludge: The Polish perspective. Renew. Energy 2010, 35, 1914–1919.
4. Werle, S. Sewage sludge gasification: Theoretical and experimental investigation. Environ. Prot. Eng. 2013, 39, 25–32.
5. Nilsson, S.; Gómez-Barea, A.; Cano, D.F. Gasification reactivity of char from dried sewage sludge in a fluidized bed. Fuel 2012, 92, 346–353.
6. Werle, S.; Dudziak, M. Gaseous Fuel Production from Granular Sewage Sludge via Air Gasification—Influence of the Waste Treatment Configuration on Sludge Properties and Producer Gas Parameters. Proceedings of the 8th Dubrovnik Conference on Sustainable Development of Energy, Water and Environment Systems, Dubrovnik, Croatia, 22–27 September 2013; pp. 1–10.
7. Phuphuakrat, T.; Nipattummakul, N.; Namioka, T.; Kerdsuwan, S.; Yoshikawa, K. Characterization of tar content in the syngas produced in a downdraft type fixed bed gasification system from dried sewage sludge. Fuel 2010, 89, 2278–2284.
8. Xu, Z.R.; Zhu, W.; Li, M. Influence of moisture content on the direct gasification of dewatered sludge via supercritical water. Int. J. Hydrog. Energy 2012, 37, 6527–3535.
9. Xu, Z.R.; Zhu, W.; Li, M.; Zhang, H.W.; Gong, M. Quantitative analysis of polycyclic aromatic hydrocarbons in solid residues from supercritical water gasification of wet sewage sludge. Appl. Energy 2013, 102, 476–483.
10. Aznar, M.; San Anselmo, M.; Manyà, J.J.; Murillo, M.B. Experimental study examining the evolution of nitrogen compounds during the gasification of dried sewage sludge. Energy Fuels 2009, 23, 3236–3245.
11. Berset, J.D.; Holzer, R. Quantitative determination of polycyclic aromatic hydrocarbons, polychlorinated biphenyls and organochlorine pesticides in sewage sludges using supercritical fluid extraction and mass spectrometric detection. J. Chromatogr. A. 1999, 852, 545–558.
12. Marrero, T.W.; McAuley, B.P.; Sutterlin, W.R.; Morris, J.S.; Manahan, S.E. Fate of heavy metals and radioactive metals in gasification of sewage sludge. Waste Manag. 2004, 24, 193–198.
13. Li, L.; Xu, Z.R.; Zhang, C.; Bao, J.; Dai, X. Quantitative evaluation of heavy metals in solid residues from sub- and super-critical water gasification of sewage sludge. Bioresour. Technol. 2012, 121, 169–175.
14. Abad, E.; Martínez, K.; Planas, C.; Palacios, O.; Caixach, J.; Rivera, J. Priority organic pollutant assessment of sludges for agricultural purposes. Chemosphere 2005, 61, 1358–1369.

15. Fuentes, A.; Lloréns, M.; Sáez, J.; Isabel Aguilar, M.; Ortuño, J.F.; Meseguer, V.F. Comparative study of six different sludges by sequential speciation of heavy metals. Bioresour. Technol. 2008, 99, 517–525.
16. Szymański, K.; Janowska, B.; Jastrzębski, P. Heavy metal compounds in wastewater and sewage sludge. Annu. Set Environ. Prot. 2011, 13, 83–100.
17. Cai, Q.-Y.; Mo, C.-H.; Wu, Q.-T.; Zeng, Q.-Y.; Katsoyiannis, A. Concentration and speciation of heavy metals in six different sewage sludge-composts. J. Hazard. Mater. 2007, 147, 1063–1072.
18. Karvelas, M.; Katsoyiannis, A.; Samara, C. Occurrence and fate of heavy metals in the wastewater treatment process. Chemosphere 2003, 53, 1201–1210.
19. Werle, S.; Dudziak, M. Evaluation of toxicity of sewage sludge and gasification waste-products. Przem. Chem. 2013, 92, 1350–1353, in Polish.
20. Werle, S.; Dudziak, M. Impact of Wastewater Treatment Processes and the Method of Sludge Treatment on Combustible Properties of Sludge Derived Fuel and Gas from Gasification. Proceedings of the 40th International Conference of Slovak Society of Chemical Engineering (SSCHE), Tatranské Matliare, Slovakia, 27–31 May 2013.
21. Werle, S.; Dudziak, M. Influence of Wastewater Treatment Processes and the Method of Sludge Treatment on Sludge-Derived Fuel Properties and Gasification Gas Parameters. Proceedings of the 26th International Conference on Efficiency, Cost, Optimization, Simulation and Environmental Impact of Energy Systems, Guilin, China, 16–19 July 2013.
22. Werle, S. Influence of sewage sludge properties on the thermal management processes. Zesz. Naukowe Uniw. Zielonogórs 2013, 151, 106–112, in Polish.
23. Solid Biofuels—Methods for Moisture Determining Using Drier Method. Part 3—Moisture Analysis in General Sample; Standards PN-EN 14774-3:2010. Polish Committee for Standardization: Warsaw, Poland, 2010.
24. Solid Recovered Fuels—Determination of Volatile Content; Standards PN-EN 15402:2011. Polish Committee for Standardization: Warsaw, Poland, 2011.
25. Solid Recovered Fuels—Determination of Ash Content; Standards PN-EN 15403:2011. Polish Committee for Standardization: Warsaw, Poland, 2011.
26. Tomza-Marciniak, A.; Witczak, A. Distribution of endocrine-disrupting pesticides in water and fish from the Oder river, Poland. Acta Ichthyol. Piscat. 2010, 40, 1–9.
27. San Miguel, G.; Dominguez, M.P.; Hernandez, M.; Sanz-Perez, F. Characterization and potential applications of solid particle produced at a biomass gasification plant. Biomass Bioenergy 2012, 47, 134–144.
28. Sinag, A.; Kruse, A.; Schwarzkopf, V. Key compounds of the hydropyrolysis of glucose in supercritical water in the presence of K2CO3. Ind. Eng. Chem. Res. 2003, 42, 3516–3521.
29. Kruse, A.; Krupka, A.; Schwarzkopf, V.; Gamard, C.; Hanningsen, T. Influence of proteins on the hydrothermal gasification and liquefaction of biomass. 1. Comparison of different feedstocks. Ind. Eng. Chem. Res. 2005, 44, 3013–3020.
30. Zhang, L.H.; Xu, C.B.; Champagne, P. Energy recovery from secondary pulp/paper mill sludge and sewage sludge with supercritical water treatment. Bioresour. Technol. 2010, 101, 2713–2721.

31. Morf, P.; Hasler, P.; Nussbaumer, T. Mechanisms and kinetics of homogeneous secondary reactions of tar from continuous pyrolysis of wood chips. Fuel 2002, 81, 843–853.
32. Zhang, B.P.; Xiong, S.J.; Xiao, B.; Yu, D.K.; Jia, X.Y. Mechanism of wet sludge pyrolysis in a tubular furnace. Int. J. Hydrog. Energy 2011, 36, 355–363.
33. Zhu, W.; Xu, Z.R.; Li, L.; He, C. The behavior of phosphorus in sub- and supercritical water gasification of sewage sludge. Chem. Eng. J. 2011, 171, 190–196.
34. The Polish Ordinance of Ministry for the Environment from 24 July 2006 about on the Conditions Which Must Be Fulfilled While Discharging Waste Water to Water or Ground and on Substances Particularly Dangerous for The Water Environment; Journal of Laws No 137 Item 984. Ministry of the Environment: Warsaw, Poland, 2006.
35. Yang, C.; Wang, J.; Lei, M.; Xie, G.; Zeng, G.; Luo, S. Biosorption of zinc(II) from aqueous solution by dried activated sludge. J. Environ. Sci. 2010, 22, 675–680.
36. Thawornchaisit, U.; Pakulanon, K. Application of dried sewage sludge as phenol biosorbent. Bioresour. Technol. 2007, 98, 140–144.

PART III

PYROLYSIS

CHAPTER 7

Pyrolysis of Waste Plastics and Whole Combustible Components Separated From Municipal Solid Wastes: Comparison of Products and Emissions

L. ZHAO, D. CHEN, Z. WANG, X. MA, AND G. ZHOU

7.1 INTRODUCTION

Landfill is the main technology to dispose solid waste in China, but lack of space for new landfill has become a growing problem especially for large urban areas. One approach to prolong current landfill life span is to excavate the aged MSW and separate the combustibles for thermal treatments, for the combustibles (upper siftings) constituted a major portion(45-62%) of the aged MSW on a volume basis (Wenjing Cui, 2007). Due to the panic of emissions of dioxins, heavy metals, etc, non-incineration methods such as pyrolysis and gasification are paid more and more attentions.

Pyrolysis produces high calorific value fuels with lower emissions in the process, the produced fuels could be easily used for power genera-

Zhao L, Chen D, Wang Z, Ma X, and Zhou G. "Pyrolysis of Waste Plastics And Whole Combustible Components Separated From Municipal Solid Wastes: Comparison of Products and Emissions." Sardinia 2011, Thirteenth International Waste Management and Landfill Symposium.

tion and heat supply, especially as pyrolysis process carried out for waste plastics, high quality oil can be obtained. In the pyrolysis process for the combustibles or waste plastics, the materials are undergone calcination in the furnace without air supply to be decomposed into gas, oil and char, the gas is recycled as fuel gas to heat the furnace. Both the oil and the char produced can be used as fuel for power generation or heat supply.

Upper siftings of aged MSW consist of plastics, fibers, rubber, very little paper, wood and bamboo sticks etc. However materials other than plastics have only a limited contribution to the final oil production during pyrolysis process. For example waste paper can not be completely cracked and its pyrolysis products at 425-515°C only contains less than 2% of oil and 6% of gas products (Wu et al. 2003). Pyrolysis of mixed wood wastes at 450-500°C generated about 66 wt% liquid products and around 11 wt% gas products (Yaman, 2004), but liquid products were poor in qulality. Pyrolysis of waste textile at around 500°C produced around 26.7 wt% light liquid hydrocarbons and 37.4 wt% of heavy liquid hydrocarbons, and 10.7 wt% of gas (Miranda et al. 2007). Pyrolysis of waste rubber would yield 65 wt% of oil and 5 wt% of gas product at 500°C (Kaminsky and Mennerich, 2001; Stelmachowski, 2009). However, pyrolysis of those materials other than plastics needs energy supply, at the same time they may cause pollutants in the pyrolysis products.

In this paper pyrolysis process for the separated waste plastics and for the upper siftings of aged MSW were compared by checking the yield and quality of the products to improve the pyrolysis process.

7.2 MATERIALS AND EXPERIMENTS

7.2.1 COMPONENTS OF THE UPPER SIFTINGS

In this research pyrolysis was carried with for two kinds of materials: the upper siftings from aged MSW that was excavated from a 7-year old landfill cell and waste plastics separated from these upper siftings. The upper siftings were mainly combustibles and their components were listed in Table 1. The composition of the plastics separated from the upper siftings was shown in Table 2.

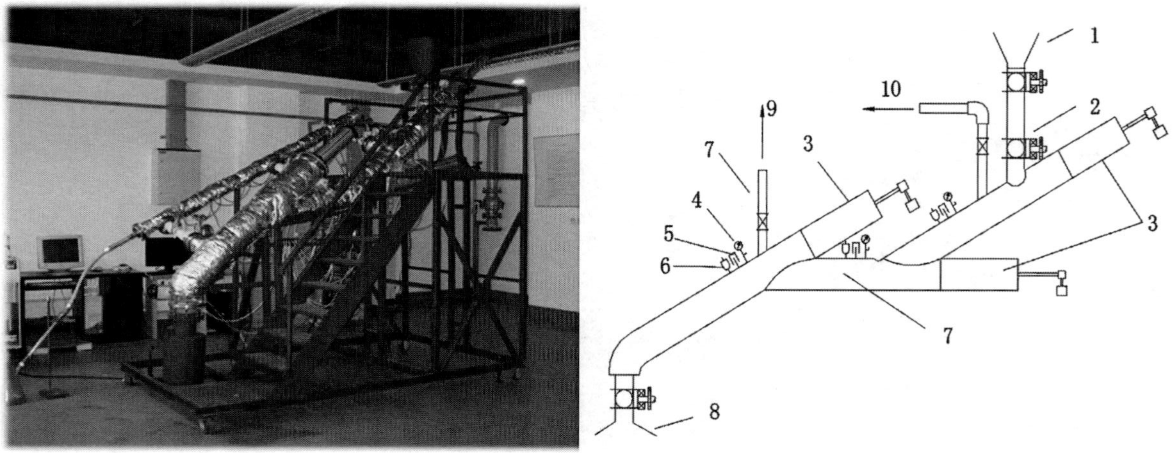

FIGURE 1: Three-section tube-furnace pyrolysis reactor (1. hopper; 2. chute valve; 3. air-driven pusher; 4.thermometer; 5. manometer; 6.security exit; 7.tube furnace; 8.char tank; 9. gas and oil products; 10. HCl)

TABLE 1: Components of the upper siftings for pyrolsis disopsal from 7-year old landfill cell.

Compo-nent	Mass ratio, %	Lower heat value, MJ/kg	Moisture wt%	C wt%	H wt%	O wt%	N wt%	S wt%	Cl wt%	Impurities contained (wt%)
Plastics	69.7	31	34	77	12	-	0.30	3.3	0.43	37.01
Rubber	3.2	19	5.3	47	5.2	15	2.6	4.2	0.41	6.06
Bamboo	12.4	16	44	43	6.0	41	1.0	3.2	0.37	11.15
Textle	13.3	17	18	58	6.7	26	0.30	2.3	0.18	20.67
Paper	1.5	15	32	44	6.0	44	0.30	0.20	-	24.91

TABLE 2: Composition of the waste plastics separated from the above upper siftings.

PE	PP	PVC	PET	PS	Other plastics
65.82%	18.50%	0.18%	10.31%	1.31%	3.92%

7.2.2 EXPERIMENTAL FACILITIES

The pyrolysis tests were carried out in a lab-scale tube furnace shown in Figure 1. The collection and condensation section was the same as reported in the previous work (Zhao, et al, 2011), after gas and oil products pass through the two stage condenser, the liquid products were condensed and collected while the incondensable gas was collected by a large gas container. Solid products were collected from the outlet where an air-tight container installed, as shown in Figure 1.

7.2.3 EXPERIMENTAL PROCEDURE

The upper siftings in Table 1 and waste plastics shown in Table 2 were put into the pyrolysis reactor respectively and the final reaction temperature was 550°C. After the products were collected, the mass of oil and char were measured. Then the oil were analyzed with a Shimadzu GC-

MS-QP2010 gas chromatograph mass spectrometer. Before analysis the moisture was removed. The gas products were analyzed with a calibrated GC. Then heat values of gas, oil and char products were measured with a XRY-1 type calorimeter. The impurities such as heavy metals, S, Cl and gasous polluants were also measured. For heavy metals an Optima 2100 DV type ICP-OES was used after oil and char were digested accroding to the method defined in EPA Method 3052 and 3050B respectively; and Hg was measured with a Milestone's DMA-80 direct mercury analyzer, H_2S and NH_3 in the gas were analysed with VRAE PGM-7800 & 7840 multi gas monitor.

7.3 RESULTS AND DISCUSSION

7.3.1 MASS DISTRIBUTION

The mass distribution is calculated by:

$$p_i(\%) = \frac{m_i}{(\sum_{i=1}^{3} m_i)} \times 100$$

Where p_i and m_i are the percent and mass of individual pyrolysis products respectively, of which the liquid and solid products were weighted and the weight of gas is calculated by mass balance. The results are listed as Table 4. It can be seen that the target product, namely liquid product from separated plastics is almost double of that from upper siftings, and the un-expected solid product is much lower.

TABLE 4: Mass distribution products

Phase	From upper siftings (wt%)	From plastics (wt%)
Gas	9.2	7.4
Liquid	42.7	84.4
Solid	48.0	8.2

TABLE 5: Composition of gas products from pyrolysis of upper siftings and plastics.

Gas products	From upper siftings (v/v)	From plastics (v/v)
H_2	19.01	9.53
CO	13.28	5.53
CH_4	10.20	12.80
C_2H_4	6.66	13.66
C_2H_6	6.22	8.86
C_3H_6	6.49	13.91
C_3H_8	2.32	4.79
i-C_4H_{10}	0.18	0.20
n-C_4H_{10}	2.85	1.68
CO_2	6.73	2.12
O_2	1.61	0.87
N_2	3.68	3.01
Others	12.66	14.03
Lower heat value(MJ/Nm^3)	33.4	52.1

7.3.2 GAS PRODUCTS

Non-condensable gas products were analyzed and quantified by a calibrated gas chromatography using TCD and FID detectors. Compositions of non-condensed products at 20°C are shown in Table 5.

Paper, textile and yard wastes show a high oxygen content, which increases the CO and CO_2 content and decrease the H_2 content in their pyrolysis gases, also the gas product from upper siftings shows a lower lower heating value (LHV). At the same time, gas products from plastics tend to be "heavier", a possible reason may be that the soil contained in the upper siftings acted as catalyst and enhanced the production of light gases.

7.3.3 LIQUID PRODUCTS

The moisture content in liquid products was analyzed according to the standard GB260-77 "Petroleum products-Determination of water", back flow solvent was benzene. The oil composition was analyzed according to the method ASTMD2887 "Standard Test Method for Boiling Range Distribution of Petroleum". The results were listed in Table 6. It can be seen that the harmful moisture content is much lower in the liquid products from pyrolysis of separated plastics.

TABLE 6: Liquid products of pyrolysis of upper shiftings and plastics.

Content	From upper siftings (%)	From plastics (%)
Moisture	13.3	2.3
Gasoline	27.1	24.2
Diesel	30.0	31.1
Heavy oil	29.6	42.4

The organic S contents in the gasoline products were tested by a CA-100 Gas absorption unit combined with a Chromeleon ICS-1000 analyzer. Phosphor and metal contents such as Cu, Pb, Mn and Fe etc were also analyzed. Their results were compared with GWKB1-1999 (The hazardous materials control standard for motor vehicle gasoline), as seen in Table 7.

TABLE 7: pollutants in gasoline products in the pyrolysis oil and car-use gasoline limits.

Item	From upper siftings	From plastics	GWKB1-1999
S content(mg/kg)	632.8	195.3	800
Pb (mg/L)	16.4	6.9	13
Cu (mg/L)	5.55	<0.01	1
Mn (mg/L)	<0.01	<0.01	18
Fe (mg/L)	41.9	2.1	5
P (mg/L)	5.31	3.50	1.3
Hg (mg/kg)	1.489	0.597	-

- Not mentioned in the standard.

The results in Table 7 show that the gasoline products from pyrolysis of upper siftings in aged MSW didn't meet the requirements by GWKB1-1999 except for Mn. Thus, further treatment is needed to purify the oil for being used as gasoline. While gasoline from pyrolysis of plastics produced could well meet the requirement in GWKB1-1999 except for its P content. The chemical composition of pyrolysis oil from upper siftings and plastics are listed in Table 8.

TABLE 8: Components of pyrolysis oil from upper siftings and wastes plastics.

Component	From upper siftings	From plastics
Aromatics	16.90%	19.70%
Alkanes	31.20%	27.70%
Cycloalkanes	3.80%	4.70%
Alkenes	33.10%	39.50%
S-Compounds	0.60%	0.10%
Compounds contained O	11.20%	5.70%
Others	3.10%	1.00%

From Table 8 it can be seen that oil from upper siftings contains more S-Compounds and compunds contained O, which is of worse quality compared to oil from plastics.

7.3.3 SOLID PRODUCTS

The proximate analysis of pyrolysis solid residues (chars) was carried out on according to GB212-2008, the standard proximate analysis method of coal, the results are shown in Table 9. To investigate the application potential the chars were digested according to EPA method 3050B for their heavy metal analysis; and their leaching toxicity was also checked according to the method defined in HJ/T 299, the results were shown in Table 10.

TABLE 9: The proximate analysis of pyrolysis chars.

Analysis item	From upper siftings	From plastics
Water content (%)	2.1	1.5
Ash Content (%)	82.7	58.3
Volatile content (%)	10.6	36.2
Fixed carbon (%)	4.6	4.0
Lower Heat value(kJ/kg)	3520	8920

TABLE 10: Heavy metal contents of the solid residue (char).

Element	From upper siftings (mg/kg)		From plastics (mg/kg)		Heavy metals' content in Coal* (mg/kg)	GB5085.3-2007 (mg/L)
	Contents	Leaching	Contents	Leaching		
As	36.0	0.060	14.0	0.011	-	5
Zn	1600.0	6.553	970.0	1.01	36.7-422.4	100
Pb	510.0	0.378	590.0	0.092	10.1-234.7	5
Bi	4.8	0.002	<0.001	<0.001	-	-
Co	12.0	0.022	6.5	0.004	-	-
Cd	12.0	<0.001	<0.001	<0.001	0.29-4.46	1
Ni	44.0	0.055	13.0	0.010	17.3-124.5	5
Ba	570.0	0.364	140.0	0.125	-	100
Fe	4300.0	0.266	3500.0	0.042	-	-
Mn	910.0	15.17	86.0	2.39	10.2-283.8	-
Cr	110.0	0.006	80.0	0.001	33-130	15
V	17.0	0.000	2.3	0.000	-	-
Cu	400.0	0.197	180.0	0.037	23.5-97.4	100
Sr	150.0	3.852	150.0	0.882	-	-
Be	0.4	0.002	<0.001	<0.001	-	0.02
Se	17.0	0.184	2.5	0.028	-	1
Hg	0.0	<0.001	0.0	<0.001	-	0.1

**Yang Xiaoyan et al. (2009)*

Data in Table 9 and Table 10 show that residue from pyrolysis of upper siftings has a lower LHV and higher ash content; the possible final disposal of this residue could be landfill or cement production. While the char from plastics pyrolysis has a much higher LHV, much lower heavy metal contents than residue from upper siftings, and the leaching toxicity of char from plastics is also much safer, all of these indicate that a more favorable option for pyrolysis process is to run with separated plastics. The results in Table 10 also show that the solid residues from both pyrolysis processes meet all the requirements for landfill and at least landfill provides a final solution for the residues if no better solution is available.

7.5 CONCLUSIONS

The pyrolysis tests were carried out for the whole combustibles and for the separated waste plastics from aged MSW at lab scale. The results proved that pyrolysis process for separated plastics was preferred to pyrolysis process for the whole upper siftings due to the following facts:

1. The expected liquid product from separated plastics is twice of that from upper siftings, and the un-expected solid product is much lower for plastics. Furthermore gaseous product from the plastics pyrolysis has a much higher LHV.
2. Oil product from the plastics' pyrolysis was of better quality than that from upper siftings'pyrolysis for its undesirable components such as moisture, S-compounds, harmful elements such as S, Pb, Cu, P and iron, etc were much lower; gasoline part of the oil product from the upper siftings' is far from the requirements for commercial use; while gasoline oil from waste plastics pyrolysis can meet almost all the requirements except for P concentration; which can be decreased by picking out special kind of plastics.
3. The residue from the upper siftings' pyrolysis corresponded to a much lower LHV and much higher heavy metal contents such as Mn, Cu and Cd, which would limit its utilization as fuel in boilers and kilns; while char from plastics' pyrolysis had a higher LHV and was much smaller in generation, therefore it may have the pos-

sibility to be used in cement kilns or even boilers. For chars from the both pyrolysis processes, leaching tests showed that they are safe to be disposed in sanitary landfills.

REFERENCES

1. Wenjing Cui (2007). Study on the production of RDF from excavated aged municipal solid wastes. Master Thesis, Tongji University, 2007:24L.
2. Al-Salem, S.M., Lettieri, P. and Baeyens, J. (2009) Recycling and recovery routes of plastic solid waste (PSW): A review. Waste Management In Press, Corrected Proof.
3. Wu, C.H., Chang, C.Y., Tseng, C.H. and Lin, J.P. (2003) Pyrolysis product distribution of waste newspaper in MSW. Journal of Analytical and Applied Pyrolysis vol. 67, n.1, pp. 41-53.
4. Yaman, S. (2004) Pyrolysis of biomass to produce fuels and chemical feedstocks. Energy Conversion and Management. vol. 45, n. 5, pp. 651-671.
5. Miranda, R., Sosa_Blanco, C., Bustos-Martınez, D. and Vasile, C. (2007) Pyrolysis of textile wastes: I. Kinetics and yields. Journal of Analytical and Applied Pyrolysis vol. 80, n.2, pp. 489-495.
6. Kaminsky, W. and Mennerich, C. (2001) Pyrolysis of synthetic tire rubber in a fluidised-bed reactor to yield 1,3-butadiene, styrene and carbon black. Journal of Analytical and Applied Pyrolysis vol.58-59, pp. 803-811.
7. Stelmachowski, M. (2009) Conversion of waste rubber to the mixture of hydrocarbons in the reactor with molten metal. Energy Conversion and Management vol. 50, n.7, pp.1739-1745.
8. Yang Xiaoyan, Liuxiaozhen, Du Xiang and Wang Yuehua (2009) Distribution of heavy metal element of coal powder in Jiangxi. Environmentatl Science & Technology. vol. 32, n. 001, pp. 115-117.

CHAPTER 8

Thermal and Catalytic Pyrolysis of Plastic Waste

DÉBORA ALMEIDA AND MARIA DE FÁTIMA MARQUES

8.1 INTRODUCTION

Plastics are materials that offer a fundamental contribution to our society, due to its versatility and relatively low cost. As a result of this contribution, a large amount of plastic waste is generated due to the increase in its production each year. This increase in the amount of waste does cause some environmental problems, since plastics do not degrade quickly and can remain in the environment for a long time [1-5]. A large part of this waste is disposed of in landfills or is incinerated [6,7].

However, the plastic waste are bulkier than other organic residues and thus occupy massive space in landfills and therefore the proper disposal and incineration have high costs. Furthermore, incineration of these waste plastics results in environmental problems due to increased emission of harmful compounds [2,6-8].

It is necessary for more sustainable solutions that incineration and disposal in landfills are researched and developed [4]. Thus, much research in the area of recycling and reuse of these post-consumed polymers have been carried out in order to produce raw materials and energy [1,3,7].

The various types of recycling are good options to control the increase of plastic waste, because they are environmentally friendly when compared with incineration and disposal in landfills. In fact, from recycling it is possible to recover raw materials, energy and fuel while minimizing the consumption of natural resources and raw materials. When these products and energy are recovered, the environmental impacts of industrial activity are reduced [1,3,9,10].

Municipal waste plastics are heterogeneous, unlike industrial. For homogeneous plastic waste, the repelletization and remoulding can be a simple and effective means of recycling. However, when these wastes are heterogeneous and consist of mixtures of resins, they are unsuitable for such recovery. In this case, other forms of recycling [11] are necessary. Each recycling method provides a number of advantages that make them beneficial for local and specific applications [12]. Appropriate treatment of plastic waste is an important question for waste management, due to energy, environmental, economic, and political [11] aspects.

The plastics recycling methods, in accordance with ASTM D5033-00, are divided into four types according to the final result, one of them being the tertiary or chemical recycling. In this type of recycling chemical degradation leads to production of liquid fuels and chemicals with high added value from waste plastic fragments or segregated [2,8,13,14].

One of the tertiary recycling methods is pyrolysis. This process can be thermal or catalytic and is a promising alternative that allows the conversion of polymers into gas and liquid hydrocarbons [4,15,16].

Pyrolysis is a process with relatively low cost from which a wide distribution of products can be obtained. In the process of pyrolysis, where heating occurs in the absence of oxygen, the organic compounds are decomposed generating gaseous and liquid products, which can be used as fuels and/or sources of chemicals. Meanwhile, the inorganic material, free of organic matter, remains unchanged under the solid fraction and can be recycled later [17].

The thermal pyrolysis requires high temperatures, which often results in products with low quality, making this process unfeasible. This occurs because the uncatalyzed thermal degradation gives rise to low molecular weight substances, however in a very wide range of products [13,15,16].

This method can be improved by the addition of catalysts, which will reduce the temperature and reaction time and allow the production of hydrocarbons with a higher added value, such as fuel oils and petrochemical feedstocks [4,11,18-21]. That is, the use of catalysts gives an added value to the pyrolysis and cracking efficiency of these catalysts depends both on its chemical and physical characteristics. These particular properties, promote the breaking of C-C bonds and determine the length of the chains of the products obtained [17,22].

For Brazilian cities, the percentage of high and low density polyethylene (HDPE and LDPE, respectively), polyethylene terephthalate (PET), Poly(vinyl chloride) (PVC) and polypropylene (PP) found in municipal solid wastes are 89% and the other polymers account for the other 11% [13]. Therefore, polyolefins (PE, PP and their copolymers) are the most widely used thermoplastics for several applications and are most of the polymeric residues, that make up 60-70% of municipal solid waste [23].

Tertiary recycling of plastic waste containing PVC releases hydrogen chloride, which causes corrosion of the pyrolysis reactor and formation of organochlorine compounds [23]. The presence of chlorine is very harmful for use as fuel in the pyrolysis liquid products obtained [24]. Although plastic waste may be considered economical sources of chemicals and energy, recycling of mixed plastic waste containing PVC not only result in the formation of volatile organic compounds in products, also in the emission of pollution when they are applied [23].

Moreover, PET may be mechanically recycled obtaining fibers for carpets, clothes and bottles. The products obtained in this recycling are of high quality that can be compared with virgin polymer [12]. Therefore, PET and other special polymers should be removed from municipal waste by mechanical recovery, which is economically viable.

8.1.1 PYROLYSIS

The tertiary or chemical recycling includes a variety of processes that enable the generation of high value products such as fuel or chemicals [11,16,19-21,25-27].

In this method, the plastic waste is processed to produce basic petrochemical compounds, which can be used as raw material for new plastics. This process has the advantage of working with mixed and contaminated plastics [12,18,20,27].

Recently, much attention has been directed to chemical recycling, particularly the uncatalyzed thermal cracking (thermolysis), catalytic cracking and steam decomposition, as methods for producing various hydrocarbon fractions in the range of fuel, from solid waste plastics [12].

In the case of polymers, pyrolysis stands out as tertiary recycling method, however this cracking gives rise to low molecular weight substances, however unfortunately in a very wide range of products, in the case of non-catalyzed thermal decomposition [11,13,15,16,18,26]. The pyrolysis can be carried out at different temperatures, reaction times, pressures, in the presence or absence of catalysts and reactive gases. The pyrolysis process involves the breaking of bonds, and is generally endothermic and hence the supply of heat is essential to react the material[28]. In polymeric samples, the decomposition process may occur through the elimination of small molecules, chain scission (depolymerization) or random cleavage [29].

In the pyrolysis process, the sample is heated in the absence of oxygen and the organic compounds are decomposed generating gaseous and liquid products. On the other hand, the inorganic part of the sample, free from organic matter remains practically unchanged in the solid fraction enabling their separation and recovery for subsequent reuse. Therefore, the pyrolysis is an attractive alternative technique for recycling waste plastics recycling [2,8,17,24,30].

Thermal pyrolysis involves the decomposition of polymeric materials by means of temperature when it is applied under inert atmospheric conditions. This process is usually conducted at temperatures between 350 and 900 °C. In the case of polyolefins, which make up much of urban waste plastics, the process proceeds through random cleavage mechanism that generates a heterogeneous mixture of linear paraffins and olefins in a wide range of molar masses [11,18,20,21].

On the other hand, the catalyzed pyrolysis promotes these decomposition reactions at lower temperatures and shorter times, because of the presence of catalysts that assist in the process. Thus, the catalytic pyrolysis presents a number of advantages over thermal, such as lower energy consumption and product formation with narrower distribution of the number of carbon atoms, which may be directed to aromatic hydrocarbons with light and high market value [11,18-21,26].

The kinetics of degradation and the pyrolysis mechanism are still being studied and discussed. Degradation has a very complex mechanism, so adequate description of decomposing a mixture of polymers is difficult, even more so in the presence of catalysts and a process with several stages [7,30]. In order to solve this problem there are some methods based on the mass loss curve during pyrolysis.

Thermogravimetric analysis (TGA) is a method that can be used to determine the loss of mass and kinetic parameters. Thermogravimetric analysis of pyrolysis involves the thermal degradation of the sample in an inert atmosphere obtaining simultaneously the weight loss values of the samples with increasing temperature at a constant heating rate [4,21,31].

Most of the techniques that are used to monitor the reactions, both for the identification of products of the gas phase and by thermogravimetric analysis, will only detect the reaction when the molecules of the products become small enough to evaporate in the gas fraction and can be observed as gas fraction or by means of mass loss of the initial sample. Is possible to follow the reaction from the beginning, since each broken link consume certain amount of energy. Thus, by measuring the heat flow into the sample during the reaction (using for example the calorimeter DSC method), it is possible to measure the rate of broken bonds occurring in the sample [4].

The reaction rates and other kinetic parameters of the degradation of the polymer are dependent on the chemical structure of these polymers. Generally the CC bonds of the polymer backbone are broken forming a higher degree of branching structures, due to the lower thermal stability of the tertiary carbon atom. Moreover, the mechanism may also be affected by contaminants. The actual reason for the differences between the rates of degradation of the macromolecules has been explained by the distortion of electron density from the degraded polymer, which depends primarily on the side group linked to the main chain of the macromolecule. For this reason, polypropylene (PP) is less stable than polyethylene (LDPE, HDPE or LLDPE), for example [7].

The mechanism of degradation of polymers has generally been described as free radical in the case of a thermal process without catalyst. However, when catalysts are used, it is generally ionic mechanism [7].

When catalysts are utilized in the pyrolysis occur two kinds of decomposition mechanisms simultaneously: thermal cracking, which in turn can follow different mechanisms (random chain scission, scission the end of the chain and/or elimination of side groups) and catalytic cracking (carbenium ions adsorbed on the catalyst surface, beta scission and desorption). As a result, a wide variety of products is generated, which in turn will react with each other resulting in a countless number of possible reaction mechanisms [30].

For the pyrolysis of polyolefins, the degradation mechanism occurs by random chain scission, where free radicals are generated propagating chain reactions and thus resulting in the cracking of polymers in a wide range of hydrocarbons that make up liquid and gaseous fractions[32]. Several factors influence the process and the most important are: residence time, temperature and the type of pyrolysis agent. When the residence time and temperature increase, the composition of the obtained product shifts to more thermodynamically stable compounds [2,8,20,32].

The pyrolysis products can be used as an alternative fuel or as a source of chemicals [30]. The composition of the product also depends on the presence of catalysts (including concentrations and types). Higher temperatures decrease the yield of hydrogen, methane, acetylene and aromatic compounds, whereas lower temperatures favor the generation of gas products [32].

Previous experiments to evaluate the polymer degradation process are important because they provide information on the feasibility of recycling these polymers raw materials and even fuels. However, most studies are focused on pyrolysis of pure polymers and unmixed [7].

8.1.1.1 THERMAL PYROLYSIS

The pyrolysis of waste plastics involves the thermal decomposition in the absence of oxygen/air. During the pyrolysis, the polymer materials are heated to high temperatures and thus, their macromolecules are broken into smaller molecules, resulting in the formation of a wide range hydrocarbons. The products obtained from the pyrolysis can be divided into non-condensable gas fraction, liquid fraction (consisting of paraffins, olefins, naphthenes and aromatics) and solid waste. From the liquid fraction can be recovered hydrocarbons in the gasoline range (C4-C12), diesel (C12-C23), kerosene (C10-C18) and motor oil (C23-C40) [1,3,18,20,33-35].

The thermal cracking usually produces a mixture of low value hydrocarbons having a wide variety of products, including hydrogen to coke. In general, when the pyrolysis temperature is high, there is increased production of non-condensable gaseous fraction and a lower liquid fuels such as diesel. The yield and composition of the products obtained are not controlled only by the temperature but also the duration of the reaction [33].

The thermal pyrolysis proceeds according to the radical chain reactions with hydrogen transfer steps and the gradual breakdown of the main chain. The mechanism involves the stages of initiation, propagation and/or free radical transfer followed by β chain scission and termination [20,34,36]. This mechanism provides many oligomers by hydrogen transfer from the tertiary carbon atom along the polymer chain to the radical site[18]. The thermal cracking is more difficult for the high density polyethylene (HDPE), followed by the low density (LDPE) and then by polypropylene (PP) [20]. This is due to high content of tertiary carbons of PP.

The initiation step comprises homolytic breaking of carbon-carbon bond, either by random chain scission as by cleavage at the end of the chain, resulting in two radicals [36,37]. For PP and PE the chain scission occurs at random [37].

This step is followed by hydrogen transfer reactions intra/intermolecular forming more stable radicals secondary. These intermediate radicals can be submitted to break the carbon-carbon bond by scission β to produce compounds saturated or with unsaturated terminal and new radicals. The transfer of intra/intermolecular hydrogen depend on the experimental conditions, the first of which leads to an increase in the production of olefins and diolefins, paraffins results in the second [34,36,37].

The termination reactions can occur, for example, by disproportionation, which can produce different olefins and alkanes or a combination of radicals can lead to the same products. Branched products can be formed from the interaction between two secondary radicals or between a secondary radical with a primary [36,37].

As a consequence of these mechanisms, the thermal pyrolysis leads to a wide distribution of hydrocarbon, a C5-C80 range, each fraction being mainly composed of diene, 1-olefin and n-paraffin. At high temperatures hydrogen is formed in significant amounts. Products obtained by thermal cracking are of limited commercial value, especially being applied as fuel. For heavy oils, it has been proposed its use as a wax[36]. Obtaining this wide range of products is one of the major drawbacks of this technique, which requires temperatures of 500 °C to 900 °C. These factors severely limit its applicability and increase the cost of recycling raw material of plastic waste [23].

8.1.1.2 CATALYTIC PYROLYSIS

The thermal pyrolysis requires high temperatures due to the low thermal conductivity of polymers [20], which is not very selective and a possible solution to reduce these reaction conditions is the use of catalyzed pyrolysis. Catalytic pyrolysis is an alternative to the recycling of pure or mixed plastics waste [30]. The catalyst can promote:

- decomposition reactions at low temperatures with lower energy consumption [15,20,36,38,39];
- reduced costs [40];
- increase the yield of products with higher added value [20,38,40];
- increase the process selectivity [39,41];
- faster cracking reactions, leading to smaller residence times and reactors with smaller volumes [36];
- inhibiting the formation of undesirable products [36];
- inhibiting the formation of products consisting primarily of cyclic hydrocarbons, aromatic and branched, in the case of polyolefins catalytic cracking [36];
- obtain liquid products with a lower boiling point range [33].

Homogeneous and heterogeneous catalyst systems have been employed in the cracking polymers. In general, heterogeneous catalysts have been more used due to the ease of their separation and recovery of the reaction [36,39]. The homogeneous catalysts especially used are Lewis acids, as $AlCl_3$, fused metal tetracloroaluminatos (M $(AlCl_4)$ n), where the metal may be lithium, sodium, potassium, magnesium, calcium or barium and n can be 1 or 2) [36].

A wide variety of heterogeneous catalysts has been used and among them are: conventional solid acids (such as zeolites, silica-alumina, alumina and FCC catalysts (Fluid Catalytic Cracking)), mesostructured catalysts (such as MCM-41 etc.), nanocrystalline zeolites (such as n-HZSM-5), among others [25,35,36,39].

Many studies have been carried out describing the cracking of pure polyolefins over various solid acids such as zeolites, clays, among others. The use of zeolites has been shown to be effective in improving the quality of products obtained in the pyrolysis of polyethylene and other addition polymers. The acidity of their active sites and its crystalline microporous structure (textural properties) favor hydrogen transfer reactions and thereby make them suitable for obtaining high conversions of gas at relatively low temperatures, between 350 and 500 °C [11,18,22,41-44]. That is, these features allow milder operating conditions (lower temperatures and reaction times) than a thermal pyrolysis [4,25,30,45].

Differences in the catalytic activity of these solids are related to their acidic properties, especially the strength and number of acidic sites. The properties of these solid structures, as the specific area, particle size and pore size distribution, also have a crucial role in their performance, they control accessibility of voluminous molecules of the polyolefin internal catalytically active sites. While most work on catalytic cracking of polymers has been performed with pure polymers, it is accepted that the decomposition process can be affected by the presence of contaminants as well as chemical changes that occur in the polymer structure during use [11,20,21,34,42,46].

As mentioned, the catalyst pore size and acidity are important factors in the catalytic cracking of polymers [40,43,47]. Generally, the level of catalytic activity in the polyolefin pyrolysis increases with increasing the number of acidic sites. Thus, it is known that zeolite catalysts achieve higher conversions acids non-zeolitic catalysts [42].

The mechanism of this process which involves the formation of a carbenium ion (isomerization, random chain scission and β cleavage, hydrogen transfer, oligomerization/alkylation, aromatization) is influenced by the strength, density and distribution of the acid sites of the catalyst. This determines the products obtained in these reactions. Solid acid catalysts such as zeolites, favor hydrogen transfer reactions due to the presence of many acid sites [11,18,22,36,42,44].

The acid strength of the solid is characterized by the presence of Lewis or Brønsted acid sites. In the case of crystalline solid acids, it is believed that most of the acid sites are located inside the pores of the material, as in the case of zeolites [11,42].

Cracking is processed either by random chain scission (medium or weak acidity), for scission at the end of the chain (strong acidity) to give waxes and distillates (gasoil, gasoline) or light hydrocarbons (C3-C5 olefins), respectively. These primary cracking products may be removed from the reaction medium or subjected to secondary reactions (such as oligomerization, cyclization and aromatization). The relative extent of these reactions is connected to the acidity and properties of catalyst, but also to experimental variables employed (such as reactor type, temperature, residence time, etc.) [36].

Catalysts having acidic sites on the surface and with the possibility of donating hydrogen ion increase rate of the isomerization products and increase the yield of hydrocarbon isomers and the quality of the fuel formed. Catalysts containing strong acid sites, higher density, are more effective in cracking polyolefins. However, the strong acidity and high pore size cause rapid deactivation of the catalyst. Thus, according to literature, it is preferable to carry out the pyrolysis of polyolefins in the presence of a catalyst with light acidity and long life [33].

Other types of catalysts which may be used in the pyrolysis process are catalysts with Lewis acid sites which are electron pair acceptors. As examples of such catalysts, there are $AlCl_3$, $FeCl_3$, $TiCl_4$ and $TiCl_3$, which are strong Lewis acids [47]. These catalysts may be dissolved in molten polymer, which substantially increases the cracking efficiency while reducing its consumption. These types of catalysts have acidic sites on their surfaces that change the charge distribution in the carbon chain, making them capable of abstracting hydride ions of hydrocarbons to produce carbonium ions. This increases the catalytic effect, enabling a reduction in pyrolysis temperature and promoting the generation of ions for olefinic and aromatic compounds [32].

However, the cost of the catalyst can greatly affect the economy of the process, even if it shows a good performance. To reduce this cost and make it even more attractive process, you can reuse the catalyst or use it in smaller quantities [23,42,48]. The biggest problem in the use of catalysts in the pyrolysis of plastics is that coke formation deactivates the catalyst over time, thereby decreasing its life cycle [33].

8.1.2 COMPARISON BETWEEN THERMAL AND CATALYTIC PYROLYSIS

Seo et al. [49] studied the catalytic degradation of HDPE using a batch reactor at a temperature of 450 °C. As shown in Table 1, the pyrolysis performed with the zeolite ZSM-5 had higher yield of the gaseous fraction and smaller liquid fraction when compared with thermal cracking. According to the authors, this is explained by the properties of the catalyst. Most zeolites, including ZSM-5, showed excellent catalytic efficiency in cracking, isomerization and aromatization due to its strong acidic property and its microporous crystalline structure. The ZSM-5 zeolite has a three-dimensional pore channel structure with pore size of 5.4 × 5.6 Å which allows an increased cracking of larger molecules, beyond the high Si/Al ratio which leads to an increase in thermal stability and acidity. Thus, initially degraded material on the external surface of the catalyst can be dispersed in the smaller internal cavities of the catalyst thus decomposed gaseous hydrocarbons (molecules with smaller sizes).

Marcilla et al. [34] also used a batch reactor to evaluate the thermal and catalytic pyrolysis of HDPE and LDPE with HZSM-5 catalyst. The processing temperature was 550 °C and the results are shown in Table 2. As can be seen, the condensable products were the major fraction for the thermal process and no solid fraction (coke) was detected. For the catalytic process an increase of the gas fraction, and this is due to the HZSM-5 catalyst present, which has strong and weak acid sites and an average pore size

TABLE 1: Yield in thermal and catalytic pyrolysis of HDPE with ZSM-5 [49].

Product Yield (% wt.)		Thermal Pyrolysis	Catalytic Pyrolysis
Gas Fraction		13.0	63.5
Liquid Fraction	Total	84.0	35.0
	C_6-C_{12}	56.55	99.92
	C_{13}-C_{23}	37.79	0.08
	>C23	5.66	0.0
Solid Fraction		3.0	1.5

small. As mentioned above, this facilitates cracking leading to compounds with small sizes (gas fraction).

The results for the batch reactor are similar. However, there are studies where the values for each product obtained are different. This is because in this type of reactor the heat transfer is not as favored and, consequently, other factors such as the size and quantity of the sample or the carrier gas flow can determine the type of product formed. Moreover, in such reactors the extent of secondary reactions is smaller than the fluidized bed reactor. Using fixed beds where polymer and catalyst are contacted directly leads to problems of blockage and difficulty in obtaining intimate contact over the whole reactor. Without effective contact the formation of large amounts of residue are likely, and scale-up to industrial scale is not feasible [15]. The low thermal conductivity and high viscosity of the plastic may lead to a difficulty in mass transfer and heat. These factors influence the distribution of products, in conjuction with the operating conditions [50].

8.1.3 ZEOLITES

Zeolites are microporous crystalline aluminosilicates of the elements of group 1A or 2A (especially sodium, potassium, magnesium and calcium), whose chemical composition can be represented as follows: $M_2/nO.Al_2O3.ySiO_2.wH_2O$, where y varies from 2 to 10, n is the valence of the cation and w is the amount of structural water [36]. Currently it is known the existence of minerals which have all essential requirements to be classified

TABLE 2: Yield of the thermal and catalytic pyrolysis of LDPE and HDPE with HZSM-5 [34].

Product Yield (% wt.)	LDPE	HDPE	LDPE-HZSM-5	HDPE-HZSM-5
Gas Fraction	14.6	16.3	70.7	72.6
Liquid Fraction/wax	93.1	84.7	18.3	17.3
Solid Fraction	-	-	0.5	0.7

as zeolites, however, instead of aluminum (Al) and silicon (Si) occupying the tetrahedral positions are present elements such as phosphorus (P), beryllium (Be), among others [51,52].

They are composed of tetrahedra of SiO_4, AlO_4 and PO_4 as primary structural units, which are linked through oxygen atoms. Each oxygen atom is shared by two silicon or aluminum atoms, thus giving rise to a three-dimensional microporous structure [46,53]. The combination of these two primary structures is found in the common zeolites, developing cavities of various shapes and sizes which are interconnected [42,51,53,54].

The AlO4 tetrahedron has a negative charge of -1, because the aluminum has a valence of +3, which is less than the valence of +4 silicon. This charge is balanced by cations of alkali metals or alkaline earth metals (typically Na^+, K^+, Ca^{+2} or Mg^{+2}) present inside the porous zeolite structure by means of cation exchange, may be replaced by other cations. When these cations are exchanged for protons, zeolite acid sites are formed. This exchange allows modification of the original properties of zeolites. The acidity of the zeolite can be the Brønsted acid type, proton donors or Lewis acid type, pair of electron acceptor [46,53]. These channels and cavities are occupied by ions, water molecules or other adsorbates which, due to high mobility, allow the ion exchange [51,53].

The pore size corresponding to two-dimensional opening zeolite is determined by the number of tetrahedral atoms connected in sequence. The three-dimensional interactions lead to the most different geometries, forming from large internal cavities to a series of channels crossing the whole zeolite [55].

The pores of zeolites function as molecular sieves, blocking the free diffusion of large, bulky molecules inside the internal surface of the catalyst [41,54]. These molecular sieves combine high acidity with selectivity form. That is, are selective to separate molecules according to their shape and/or size, besides having a high specific area and high thermal stability to catalyze a variety of hydrocarbon reactions, including the cracking of polyolefins. The reactivity and the selectivity of zeolites as catalysts are determined by its high number of active sites, which are caused by an imbalance of charge between the silicon and aluminum atoms in the crystal, making the zeolite of the structural unit has a charge balance total least one [42,51].

However, the process of rupture of the polymer molecules starts on the external surface of the zeolites, since the polymer chains must be broken before penetrating the internal pores of the zeolites, due to its small pore size. The zeolites have a specific pore size and the access of polymer molecules to internal reactive sites of the catalyst, as well as the final products within the pores are limited by their size. As mentioned, the catalyst pore size and acidity are important factors in the catalytic cracking of polymers [40,43,47]. Generally, the level of catalytic activity in the pyrolysis of polyolefins increases with increasing the number of acidic sites. Thus, it is known that zeolite catalysts achieve higher conversions than non-zeolitic catalysts acids [42]. In addition, branching of the polymer or end chain of polyethylene can penetrate the pores of the zeolites, reacting the acid sites located there and so increasing the activity [34].

During the catalyzed pyrolysis, the polymer melts and is dispersed around the catalyst. The molten polymer is drawn into the spaces between the particles and therefore the active sites on the external surface of the catalyst. Reactions at the surface produce a low molecular weight materials, which are sufficiently volatile at the temperature of the reaction can diffuse through the polymer film as a product or may react even more in the pores. These reactions proceed via carbocation as transition state. The reaction rate is governed both by the nature of the carbocation formed as the nature and strength of the acid sites involved in catalysis. Regardless of how the carbocation is formed, it may be subjected to any of the following methods: load isomerization, the isomerization chain, hydride transfer, transfer of alkyl groups and formation and breaking of carbon-carbon bonds. As a result of this complex procedure, the product distribution reflects the action of the catalyst, which in turn is influenced by the size of its pores and for its chemical composition [34,56].

The catalytic decomposition of the polyethylene occurs at the carbenium ion mechanism. The initial step occurs either by abstraction of the hydride ion (for Lewis acid sites) or by addition of a proton (the Brønsted acid sites) in the C-C bonds of polyethylene molecules, or by thermal decomposition of polyolefins. Successive scission of the main chain occur to produce fragments having lower molecular weights than that of polyethylene. The resulting fragments are cracked or desidrociclizados in subsequent steps [18].

The acid sites on the catalyst surface are responsible for the initiation of the carbocationic mechanism, which induces the degradation of polyethylene and polypropylene. As mentioned above, these acid sites are originated the generated load imbalance when AlO_4^- is incorporated in the structure of zeolites. The content of AlO_4^- determines the number of acid sites in the catalyst while topological factors related to its crystalline or amorphous structure influence the strength of these acidic centers. Textural characteristics control the access of molecules that are reacting in the catalytic sites. This accessibility is important in catalyzed reactions involving large molecules such as polymers [21,57].

For presenting a microporous structure, zeolites have a higher internal surface than the external surface and this enables the mass transfer between these surfaces. However, the pore size is an important factor in this transfer, because only molecules with sizes smaller than these pores can enter or leave these spaces, which vary from one to another zeolite [53].

Some chemical and physical characteristics of zeolites ensures them their catalytic capacity. Among these characteristics can be cited: high specific area and adsorption capacity; active sites (which may be acidic) whose strength and concentration can be directed to a specific application; size channels and cavities compatible with the size of many molecules and a network of canals and cavities that provides you with a selectivity of shape, selectivity to the reactant, product and transition state species [53].

One of the factors that can affect the catalytic activity of zeolites is their deactivation by coke deposition on their channels. However, this coke formation rate depends on several factors, including: structure and acidity of the pores and the reaction conditions (such as temperature, pressure and nature of the reactants) [53].

The synthetic zeolites present some advantages and disadvantages in relation to natural zeolites. Among the advantages may be mentioned the purity, uniformity in size and shape of the channels and cavities, and a predefined chemical composition. The disadvantage has been their high cost and because of this, the synthetic zeolites are mainly intended for specific applications, where there is a need for a uniform composition and structure, for example, in the petroleum cracking process. Already the natural have a greater abundance and a lower cost of production, particularly if used in its in natura or if they require little beneficiation complex processes [51].

8.2 CONCLUSIONS

Consumption of plastics has increased over the years and the concern with their waste generated too. Because of this many studies have been done with the aim to recover or recycle the waste.

Pyrolysis has been effective compared to other disposal methods, because it can reuse the energy and the raw materials contained in those waste, reducing thereby the environmental impacts caused by the inadequate disposal of these waste plastics.

The pyrolysis process may be thermal or catalytic. Thermal degradation occurs by radical mechanism, and as a result of this mechanism the products formed have a broad distribution of the number of carbon atoms in the main chain.

In this type of the endothermic process due to the low thermal conductivity of polymers, there is a need for high temperatures. Because of that there is a high expenditure of energy. In order to decrease this temperature, catalysts may be used.

With the catalytic pyrolysis, the products obtained have a more narrow distribution of the number of carbon atoms being directed to more specific products. The composition and amount of the obtained products are listed as type of catalyst used. Furthermore, the catalytic reaction decreases the degradation time and the fraction of solid waste formed.

Generally, the catalysts used in the catalytic degradation are solid acids such as zeolites. This type of degradation involves production of the intermediate carbenium ion by hydrogen transfer reactions. Zeolites used favor these reactions due to their sites acids that help in the process of breaking the polymer macromolecules. This breaking process begins on the surface of the zeolite, because the polymer needs to be broken into smaller molecules before entering the internal pores of these solids, due to the small size of their pores. Zeolites have a specific molecular pore size and access of such molecules to catalytic reactive sites, as well as growth of the final products within such pores is limited by its size.

The other experimental parameters such as temperature, reaction time, reactor type and flow of carrier gas also influence the composition of the products obtained. Pyrolysis can be carried out either for pure polymers or for polymer blends.

REFERENCES

1. Mastral, J. F., Berrueco, C., & Ceamanos, J. (2007). Theoretical prediction of product distribution of the pyrolysis of high density polyethylene. Journal of Analytical and Applied Pyrolysis, 80(2), 427-438. http://dx.doi.org/10.1016/j.jaap.2006.07.009.
2. Abbas-Abadi, M. S., Haghighi, M. N., & Yeganeh, H. (2012). The effect of temperature, catalyst, different carrier gases and stirrer on the produced transportation hydrocarbons of LLDPE degradation in a stirred reactor. Journal of Analytical and Applied Pyrolysis, 95, 198-204. http://dx.doi.org/10.1016/j.jaap.2012.02.007.
3. Arabiourrutia, M., Elordi, G., Lopez, G., Borsella, E., Bilbao, J., & Olazar, M. (2012). Characterization of the waxes obtained by the pyrolysis of polyolefin plastics in a conical spouted bed reactor. Journal of Analytical and Applied Pyrolysis, 94, 230-237. http://dx.doi.org/10.1016/j.jaap.2011.12.012.
4. Coelho, A., Costa, L., Marques, M. M., Fonseca, I. M., Lemos, M. A. N. D. A., & Lemos, F. (2012). The effect of ZSM-5 zeolite acidity on the catalytic degradation of high-density polyethylene using simultaneous DSC/TG analysis. Applied Catalysis A: General, 413-414, 183-191. http://dx.doi.org/10.1016/j.apcata.2011.11.010.
5. Abbas-Abadi, M. S., Haghighi, M. N., & Yeganeh, H. (2013). Evaluation of pyrolysis products of virgin high density polyethylene degradation using different process parameters in a stirred reactor. Fuel Processing Technology, 109, 90-95. http://dx.doi.org/10.1016/j.fuproc.2012.09.042.
6. Stelmachowski, M. (2010). Thermal conversion of waste polyolefins to the mixture by hydrocarbons in the reactor with molten metal bed. Energy Conversion and Management, 51(10), 2016-2020. http://dx.doi.org/10.1016/j.enconman.2010.02.035.

7. Miskolczi, N., & Nagy, R. (2012). Hydrocarbons obtained by waste plastic pyrolysis: comparative analysis of decomposition described by different kinetic models. Fuel Processing Technology, 104, 96-104. http://dx.doi.org/10.1016/j.fuproc.2012.04.031.
8. Demirbas, A. (2004). Pyrolysis of municipal of plastic wastes for recovery of gasoline-range hydrocarbons. Journal of Analytical and Applied Pyrolysis, 72(1), 97-102. http://dx.doi.org/10.1016/j.jaap.2004.03.001.
9. Valle, M. L. M., Guimarães, M. J. O. C., & Sampaio, C. M. S. (2004). Degradação de poliolefinas utilizando catalisadores zeólitas. Polímeros: Ciência e Tecnologia, 1(14), 17-21. http://dx.doi.org/10.1590/S0104-14282004000100009.
10. Shah, S. H., Khan, Z. M., Raja, I. A., Mahmood, Q., Bhatti, Z. A., Khan, J., Farooq, A., Rashid, N., & Wu, D. (2010). Low temperature conversion of plastic waste into light hydrocarbons. Journal of Hazardous Materials, 179(1-3), 15-20. http://dx.doi.org/10.1016/j.jhazmat.2010.01.134. PMid:20172649.
11. Panda, A. K., Singh, R. K., & Mishra, D. K. (2010). Thermolysis of waste plastics to liquid fuel. A suitable method for plastic waste management and manufacture of value added products: a world prospective. Renewable & Sustainable Energy Reviews, 14(1), 233-248. http://dx.doi.org/10.1016/j.rser.2009.07.005.
12. Al-Salem, S. M., Lettieri, P., & Baeyens, J. (2009). Recycling and recovery routes of plastic solid waste (PSW): a review. Waste Management (New York, N.Y.), 29(10), 2625-2643. http://dx.doi.org/10.1016/j.wasman.2009.06.004. PMid:19577459.
13. Spinacé, M. A. S., & De Paoli, M. A. (2005). A tecnologia da reciclagem de polímeros. Quimica Nova, 98(1), 65-72. http://dx.doi.org/10.1590/S0100-40422005000100014.
14. Singhabhandhu, A., & Tezuka, T. (2010). The waste-to-energy framework for integrated multi-waste utilization: waste cooking oil, waste lubricating oil, and waste plastics. Energy, 35(6), 2544-2551. http://dx.doi.org/10.1016/j.energy.2010.03.001.
15. Lin, Y.-H., & Yang, M.-H. (2008). Tertiary recycling of polyethylene waste by fluidized-bed reactions in the presence of various cracking catalysts. Journal of Analytical and Applied Pyrolysis, 83(1), 101-109. http://dx.doi.org/10.1016/j.jaap.2008.06.004.
16. Huang, W.-C., Huang, M.-S., Huang, C.-F., Chen, C.-C., & Ou, K.-L. (2010). Thermochemical conversion of polymer wastes into hydrocarbon fuels over various fluidizing cracking catalysts. Fuel, 89(9), 2305-2316. http://dx.doi.org/10.1016/j.fuel.2010.04.013.
17. López, A., De Marco, I., Caballero, B. M., Adrados, A., & Laresgoiti, M. F. (2011). Deactivation and regeneration of ZSM-5 zeolite in catalytic pyrolysis of plastic wastes. Waste Management (New York, N.Y.), 31(8), 1852-1858. http://dx.doi.org/10.1016/j.wasman.2011.04.004. PMid:21530221.
18. Park, D. W., Hwang, E. Y., Kim, J. R., Choi, J. K., Kim, Y. A., & Woo, H. C. (1999). Catalytic degradation of polyethylene over solid acid catalysts. Polymer Degradation & Stability, 65(2), 193-198. http://dx.doi.org/10.1016/S0141-3910(99)00004-X.
19. Lin, Y.-H., & Yang, M.-H. (2005). Catalytic reactions of post-consumer polymer waste over fluidized cracking catalysts for producing hydrocarbons. Journal of Molecular Catalysis A Chemical, 231(1-2), 113-122. http://dx.doi.org/10.1016/j.molcata.2005.01.003.

20. Achilias, D. S., Roupakias, C., Megalokonomos, P., Lappas, A. A., & Antonakou, E. V. (2007). Chemical recycling of plastic wastes made from polyethylene (LDPE and HDPE) and polypropylene (PP). Journal of Hazardous Materials, 149(3), 536-542. http://dx.doi.org/10.1016/j.jhazmat.2007.06.076. PMid:17681427.
21. Aguado, J., Serrano, D. P., San Miguel, G., Escola, J. M., & Rodríguez, J. M. (2007). Catalytic activity of zeolitic and mesostructured catalysts in the cracking of pure and waste polyolefins. Journal of Analytical and Applied Pyrolysis, 78(1), 153-161. http://dx.doi.org/10.1016/j.jaap.2006.06.004.
22. Serrano, D. P., Aguado, J., Escola, J. M., & Rodríguez, J. M. (2005). Influence of nanocrystalline HZSM-5 external surface on the catalytic cracking of polyolefins. Journal of Analytical and Applied Pyrolysis, 74(1-2), 353-360. http://dx.doi.org/10.1016/j.jaap.2004.11.037.
23. Lin, H.-T., Huang, M.-S., Luo, J.-W., Lin, L.-H., Lee, C.-M., & Ou, K.-L. (2010). Hydrocarbon fuels produced by catalytic pyrolysis of hospital plastic wastes in a fluidizing cracking process. Fuel Processing Technology, 91(11), 1355-1363. http://dx.doi.org/10.1016/j.fuproc.2010.03.016.
24. López, A., De Marco, I., Caballero, B. M., Laresgoiti, M. F., & Adrados, A. (2010). Pyrolysis of municipal wastes: influence of raw material composition. Waste Management (New York, N.Y.), 30(4), 620-627. http://dx.doi.org/10.1016/j.wasman.2009.10.014. PMid:19926462.
25. Sakata, Y., Uddin, M.A., & Muto, A. (1999). Degradation of polyethylene and polypropylene into fuel oil by using solid acid and non-acid catalysts. Journal of Analytical and Applied Pyrolysis, 51(1-2), 135-155.
26. Lee, S.Y., Yoon, J.H., Kim, J.R., & Park, D.W. (2001). Catalytic degradation of polystyrene over natural clinoptilolite zeolita. Polymer Degradation and Stability, 74(2), 297-305.
27. Hernández, M. R., Garcia, A. N., & Marcilla, A. (2005). Study of the gases obtained in thermal and catalytic flash pyrolysis of HDPE in a fluidized bed reactor. Journal of Analytical and Applied Pyrolysis, 73(2), 314-322. http://dx.doi.org/10.1016/j.jaap.2005.03.001.
28. Buekens, A. (2006). Introduction to feedstock recycling of plastics. In J. Scheirs, & W. Kaminsky (Orgs.), Feedstock recycling and pyrolysis of waste plastics (pp. 3-42). Hoboken: John Wiley & Sons.
29. Silvério, F. O., Barbosa, L. C. A., & Piló-Veloso, D. (2008). A pirólise como técnica analítica. Quimica Nova, 31(6), 1543-1552. http://dx.doi.org/10.1590/S0100-40422008000600045.
30. Lopez-Urionabarrenechea, A., De Marco, I., Caballero, B. M., Laresgoiti, M. F., & Adrados, A. (2012). Catalytic stepwise pyrolysis of packaging plastic waste. Journal of Analytical and Applied Pyrolysis, 96, 54-62. http://dx.doi.org/10.1016/j.jaap.2012.03.004.
31. Singh, S., Wu, C., & Williams, P. (2012). Pyrolysis of waste materials using TGA-MS and TGA-FTIR as complementary characterization techniques. Journal of Analytical and Applied Pyrolysis, 94, 99-107. http://dx.doi.org/10.1016/j.jaap.2011.11.011.
32. Donaj, P. J., Kaminsky, W., Buzeto, F., & Yang, W. (2012). Pyrolysis of polyolefins for increasing the yield of monomers' recovery. Waste Management (New

York, N.Y.), 32(5), 840-846. http://dx.doi.org/10.1016/j.wasman.2011.10.009. PMid:22093704.
33. Scheirs, J. (2006). Overview of commercial pyrolysis processes for waste plastics. In J. Scheirs, & W. Kaminsky (Orgs.), Feedstock recycling and pyrolysis of waste plastics (pp. 383-434). Hoboken: John Wiley & Sons.
34. Marcilla, A., Beltrán, M. I., & Navarro, R. (2009). Thermal and catalytic pyrolysis of polyethylene over HZSM5 and HUSY zeolites in a batch reactor under dynamic conditions. Applied Catalysis B: Environmental, 86(1-2), 78-86. http://dx.doi.org/10.1016/j.apcatb.2008.07.026.
35. Lee, K.-H. (2012). Effects of the types of zeolites on catalytic upgrading of pyrolysis wax oil. Journal of Analytical and Applied Pyrolysis, 94, 209-214. http://dx.doi.org/10.1016/j.jaap.2011.12.015.
36. Aguado, J., Serrano, D. P., & Escola, J. M. (2006). Catalytic upgrading of plastic wastes. In J. Scheirs, & W. Kaminsky (Orgs.), Feedstock recycling and pyrolysis of waste plastics (pp. 73-110). Hoboken: John Wiley & Sons.
37. Lee, K.-H. (2006). Thermal and catalytic degradation of waste HDPE. In J. Scheirs, & W. Kaminsky (Orgs.), Feedstock recycling and pyrolysis of waste plastics (pp. 129-160). Hoboken: John Wiley & Sons.
38. Murata, K., Brebu, M., & Sakata, Y. (2010). The effect of silica-alumina catalysts on degradation of polyolefins by a continuous flow reactor. Journal of Analytical and Applied Pyrolysis, 89(1), 30-38. http://dx.doi.org/10.1016/j.jaap.2010.05.002.
39. Liu, W., Hu, C., Yang, Y., Tong, D., Li, G., & Zhu, L. (2010). Influence of ZSM-5 zeolite on the pyrolytic intermediates from the co-pyrolysis of pubescens and LDPE. Energy Conversion and Management, 51(5), 1025-1032. http://dx.doi.org/10.1016/j.enconman.2009.12.005.
40. White, R. L. (2006). Acid-catalyzed cracking of polyolefins: primary reaction mechanism. In J. Scheirs, & W. Kaminsky (Orgs.), Feedstock recycling and pyrolysis of waste plastics (pp. 45-72). Hoboken: John Wiley & Sons.
41. Mastral, J. F., Berrueco, C., Gea, M., & Ceamanos, J. (2006). Catalytic degradation of high density polyethylene over nanocrystalline HZSM-5 zeolite. Polymer Degradation & Stability, 91(12), 3330-3338. http://dx.doi.org/10.1016/j.polymdegradstab.2006.06.009.
42. Ofoma, I. (2006). Catalytic pyrolysis of polyolefins. Atlanta: Georgia Institute of Technology.
43. Li, X., Shen, B., Guo, Q., & Gao, J. (2007). Effects of large pore zeolite additions in the catalytic pyrolysis catalyst on the light olefins production. Catalysis Today, 125(3-4), 270-277. http://dx.doi.org/10.1016/j.cattod.2007.03.021.
44. Miskolczi, N., & Bartha, L. (2008). Investigation of hydrocarbon fractions form waste plastic recycling by FTIR, GC, EDXRFS and SEC techniques. Journal of Biochemical and Biophysical Methods, 70(6), 1247-1253. http://dx.doi.org/10.1016/j.jbbm.2007.05.005. PMid:17602751.
45. Elordi, G., Olazar, M., Aguado, R., Lopez, G., Arabiourrutia, M., & Bilbao, J. (2007). Catalytic pyrolysis of high density polyethylene in a conical spouted bed reactor. Journal of Analytical and Applied Pyrolysis, 79(1-2), 450-455. http://dx.doi.org/10.1016/j.jaap.2006.11.010.

46. Manos, G. (2006). Catalytic degradation of plastic waste to fuel over microporus materials. In J. Scheirs, & W. Kaminsky (Orgs.), Feedstock recycling and pyrolysis of waste plastics (pp. 193-208). Hoboken: John Wiley & Sons.
47. Kaminsky, W., & Zorriqueta, I.-J. N. (2007). Catalytical and thermal pyrolysis of polyolefins. Journal of Analytical and Applied Pyrolysis, 79(1-2), 368-374. http://dx.doi.org/10.1016/j.jaap.2006.11.005.
48. López, A., De Marco, I., Caballero, B. M., Laresgoiti, M. F., Adrados, A., & Aranzabal, A. (2011). Catalytic pyrolysis of plastic wastes with two different types of catalysts: ZSM-5 zeolite and Red Mud. Applied Catalysis B: Environmental, 104(3-4), 211-219. http://dx.doi.org/10.1016/j.apcatb.2011.03.030.
49. Seo, Y.-H., Lee, K.-H., & Shin, D.-H. (2003). Investigation of catalytic degradation of high-density polyethylene by hydrocarbons group type analysis. Journal of Analytical and Applied Pyrolysis, 70(2), 383-398. http://dx.doi.org/10.1016/S0165-2370(02)00186-9.
50. Lin, Y.-H. (2009). Production of valuable hydrocarbons by catalytic degradation of a mixture of post-consumer plastic waste in a fluidized-bed reactor. Polymer Degradation & Stability, 94(11), 1924-1931. http://dx.doi.org/10.1016/j.polymdegradstab.2009.08.004.
51. Monte, M. B. M., & Resende, N. G. A. M. (2005). Zeolitas naturais. In A. B. Luz, & F. A. F. Lins, Rocha e minerais industriais: usos e especificações (pp. 699-720). Rio de Janeiro: CETEM.
52. Letichevsky, S. (2008). Síntese e caracterização das zeolitas mordenita, ferrierita e ZSM-5 nanocristalinas (Tese de doutorado). Pontifícia Universidade Católica do Rio de Janeiro, Rio de Janeiro.
53. Tourinho, R.R.C. (2009). Estudo da acidez de zeolitas impregnadas com platina utilizando reações de troca H/D com aromáticos e correlações lineares de energia livre (Dissertação de mestrado). Universidade Federal do Rio de Janeiro, Rio de Janeiro.
54. Aksoy, Y. Y. (2010). Characterization of two zeolites for geotechnical and geoenvironmental applications. Applied Clay Science, 50(1), 130-136. http://dx.doi.org/10.1016/j.clay.2010.07.015.
55. Braga, A. A. C., & Morgon, N. H. (2007). Descrições estruturais cristalinas de zeolitos. Quimica Nova, 30(1), 178-188. http://dx.doi.org/10.1590/S0100-40422007000100030.
56. Pinto, F., Costa, P., Gulyurtlu, I., & Cabrita, I. (1999). Pyrolysis of plastic wastes 2. Effect of catalyst on product yield. Journal of Analytical and Applied Pyrolysis, 51, 57-71.
57. Hwang, E.-Y., Kim, J.-R., Choi, J.-K.; Woo, H.-C., & Park, D.-W. (2002). Performance of acid treated natural zeolitas in catalytic degradation of polypropylene. Journal of Analytical and Applied Pyrolysis, 62(2), 351-364.

CHAPTER 9

Facile Route to Generate Fuel Oil via Catalytic Pyrolysis of Waste Polypropylene Bags: Towards Waste Management of >20 μm Plastic Bags

NEERAJ MISHRA, SUNIL PANDEY, BHUSHAN PATIL, MUKESHCHAND THUKUR, ASHMI MEWADA, MADHURI SHARON, AND MAHESHWAR SHARON

9.1 INTRODUCTION

Due to the nondegradability of PP plastics, their heavy accumulation in the environment is causing hostile effects on ecosystem including soil erosion [1]. Conventional routes to recycle PP plastics such as mechanical recycling, land filling, incineration, and chemical recycling [2] suffer from many hostile impacts such as landfill waste, clogged waterways, occupational health hazards, energy consumption, animal death, water contamination, foreign oil dependency, toxic pollution, soil degradation, costly production/recycling of plastics, and landscape litter. Additionally, these techniques have following limitations.

1. Low-conversion efficiency.
2. No valuable by products are formed; rather one form of plastic is converted to another, which has no commercial value.
3. Requiring high energy and manual efforts.
4. Heavy pollutants are generated during the process.

Keeping these cardinal issues under consideration, a facile route to convert waste plastics into high-performance fuel oil using high temperature pyrolysis is discussed in this paper. Fuel oil generated by our method was found to have all the characteristics to be used as fuel oil in factories. Since waste plastics like polypropylene (PP) contain 85% the carbon and rest is hydrogen, this makes them extremely suitable for feedstock recycling with the production of valuable hydrocarbon products. This fact can be explained with the difference in the activation energy of two polymers. PP requires lower activation energy to break the C–H bond than polyethylene (PE) because carbon chain of PP polymer contains tertiary carbon atoms which have considerably lower resistance against degradation [3]. There are several products obtained from the catalytic pyrolysis of PP which are carbon nano-materials (CNMs) of different morphologies [4–6], wax [7–9], oil [10–13] and gases [3, 14]. Most of the works, concerned with production of oils from waste PP plastics, and conversion efficiency was found to be <50%.

Our main efforts were focused on converting WPP into less nonhazardous forms by ecofriendly method like catalytic pyrolysis. In our previous paper, conversion of PP to MWCNTs was reported [15]. However, during these conversion experiments, traces of liquid hydrocarbons were also noted. Therefore, a systematic study of oil-like liquefied hydrocarbon production was undertaken. This conversion has been demonstrated earlier by a simple thermal process in which polymers are melted and broken down to smaller molecules, at high temperature, into gaseous, liquid, and solid hydrocarbons [16]. Sarker et al. in 2012 [11] have performed thermal degradation of HDPE-2 in a fiber glass reactor system at a reaction temperature between 370 and 420°C and a reaction time of 4 h using HZSM-5 molecular sieves as catalyst to obtain hydrocarbon liquid fuel. Tymoshevskyy et al. in 2009 [17] have used pyrolysis method to convert PP, PE, and polystyrene (PS) waste to fuel with trial of several catalysts. They have separated the product of catalytic cracking used in a distillation column into gas, gasoline, light oil, and heavy oil fractions. Tiwari et al. in 2009 [12] have done the catalytic degradation of Linear Low Density PE (LLDPE) using two commercial cracking catalyst-1 and cracking catalyst-2 containing 20% and 40% US-Y zeolite, respectively, with average particle size in micrometers at temperature of 600°C to produce light hydrocarbon fuel. Catalytic pyrolysis of LDPE was investigated using various fly ash-derived silica-alumina catalysts (FSAs) by Na et al. in 2006 [18]. In the present work, our trials involve simple catalyst preparation as well as cost effective pyrolysis reactor for the synthesis of liquefied hydrocarbons. Catalytic pyrolysis due to combined impact of pyrolysis and catalytic reforming is shown as a more efficient method for processing large amounts of waste plastics [19]. We have made an attempt to utilize WPP as potential precursor for synthesis of liquefied hydrocarbons using catalytic pyrolysis. The synthesis of liquefied hydrocarbons from waste PP plastics using Ni as a catalyst in reactor at pyrolysis temperature of 550°C is reported. It was observed by Na et al. in 2006 [18] and Pandian and Kamalakannan in 2012 [20], that liquefied hydrocarbon synthesized at lower temperature or below 500°C after 1-2 days turns into wax whereas oil prepared above 500°C remains as it is after long period of time. Synthesized liquefied product was characterized by GCMS. Moreover, its flash point, pour point, viscosity, specific gravity, and density were also measured.

9.2 EXPERIMENT

Waste PP was collected from Brihanmumbai Municipal Corporation (BMC) garbage disposal centre. Immediately after collection, WPP were washed, air-dried, and shredded into small pieces prior to pyrolysis in presence of nano-sized nickel as catalyst. To synthesize Ni catalyst, 10 mL of 5 mM nickel nitrate was mixed with the same volume of ethanol under constant stirring for 2 hrs [21]. One kg of shredded WPP was mixed with

10 mL of nickel (Ni) catalyst and subjected to pyrolysis as displayed in Figure 1.

Pyrolysis of waste plastic under controlled Ar (500 sccm) gas atmosphere at fixed temperature of 550°C (the ramp temperature of 10°C/min was used during this whole reaction process until the required temperature was achieved) with 1 h of dwell time in the presence of Ni as catalyst resulted in production of liquefied hydrocarbon or oil. This liquid was collected after the furnace temperature was allowed to cool down naturally once the reaction time was over in the atmosphere of 100 sccm of Ar gas (Figure 1). Due to high temperature, plastics get fragmented into monomers and reacted with catalyst and converted into the form of respective gases form which condense through condenser and are stored in the collection tank (Figure 1).

9.3 CHARACTERISATION TECHNIQUE

Gas chromatography coupled with high resolution mass spectrometer (GC-HRMS) was done by using an Agilent 7890 chromatograph with a (15 mm × 0.25 mm × 0.25 μm) glass column packed with 80–100 mesh n-octanes, Porasil C, and with a flame ionization detector (FID). In gas chromatography, helium gas at the flow rate 1.5 mL/min was used as a carrier gas with a head space injection temperature of 250°C. The oven temperature was programmed from 40°C (2 min hold) to 140°C (10 min hold) with heating ramp rate of 8°C/min. Data scan rate of the sample was kept at 0.6 sec/scan with mass scan range of 10 to 425 m/z.

Electron energy of mass spectrometer (Jeol, AccuTOF GCV) was 70 eV, and the ion source and coupling temperatures were 230 and 300°C, respectively. The ion mass spectra derived were automatically compared to TLC, MS spectral libraries. Standard solutions were analyzed to verify the identity of the peaks by retention time and provide quantitative analysis.

Elemental analysis was carried out by CHNS (O) analyser of Thermo Finnigan (FLASH EA112 series) of IIT Bombay, Mumbai.

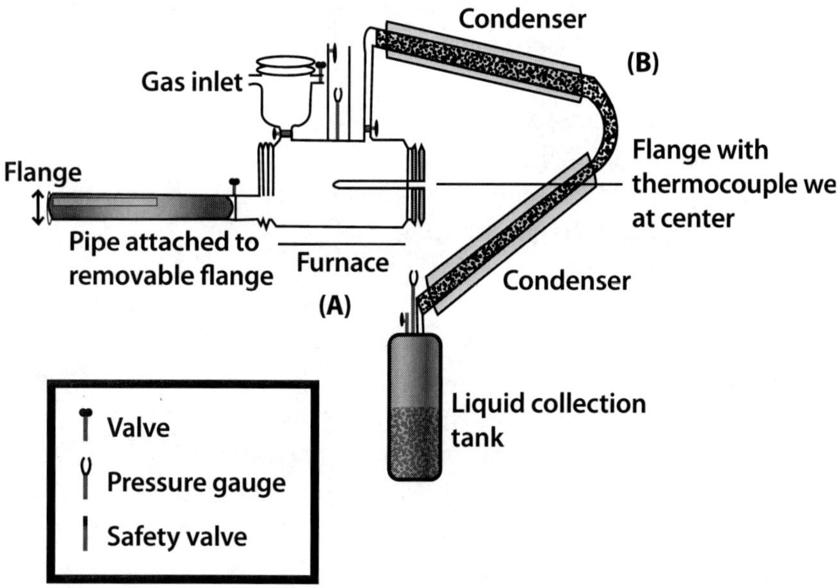

FIGURE 1: Schematic diagram of unit used for synthesizing liquefied hydrocarbon by catalytic pyrolysis reactor. In this diagram, stainless steel reactor (A), gas inlet, condenser (B), liquid collection tank, valve, pressure gauge, and safety valve have been shown

Density, API gravity, specific gravity, pour point, and flash point—all these tests of liquid hydrocarbon, were characterized by standard methods [20].

9.4 RESULT AND DISCUSSION

During the catalytic degradation of WPP, the following cardinal steps decided the efficiency of conversion.

1. Size monodispersity of the catalyst.
2. Ratio of WPP and catalyst.
3. Temperature gradient between site of pyrolysis (Figure 1(A)) and site of collection (Figure 1(B)).

As displayed in Figure 2, size of Ni catalyst was found to be ranging between 5 and 10 nm. At nanoscale, enhanced catalytic activity of Ni involved in cracking and hydrogenation leads to efficient conversion of WPP to oil [22, 23]. Additionally, efficient diffusion of Ni during the process of pyrolysis helps to accelerate the entire reaction [24].

Ratio of WPP and catalyst (1% w/w) was also found to be vital parameter due to quantity-dependent reaction mechanism of Ni catalyst. At higher quantities, Ni influences conversion of PP to higher aromatic compounds which leads to amorphous carbon, thus decreasing the yield of oil. This is also due to self-pyrolytic properties of aromatic compounds which get converted to amorphous carbon [25]. Conversion of WPP to oil mainly depends upon the vaporization followed by prompt condensation of vapours with the help of condenser. Due to unique design of our furnace, vaporisation of PP was initiated at 350°C followed by complete process at 550°C. Till the vapours reach to condenser for the formation oil, there was sharp maintenance of temperature to avoid loss of higher carbon numbers (which may decrease the yield). This gradient in the temperature of the reaction vessel (550°C) and condenser (150°C) was possible due to kink (Figure 1(B)) separating the vapours at high as well as low temperatures.

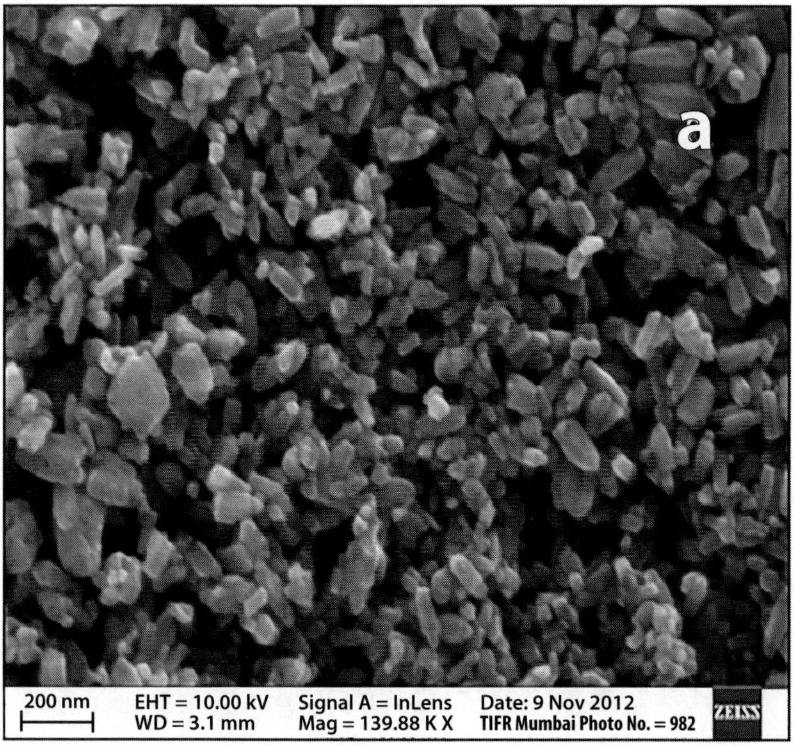

FIGURE 2: SEM and EDEX of Ni nanoparticles.

TABLE 1: Comparative study of physical properties of fuel oil obtained by WPP with other waste plastic products.

SR. no.	Test	Method of testing	Result	Waste LDPE [20]	Waste mixed plastic [12]	Waste HDPE [27]	Gulf fuel oil [28]
1	Density @ 15°C	ASTM D 4052:2002	0.7930 g/mL	0.8760 g/mL	NA	0.7828 g/mL	NA
2	Acidity (mg KOH/g)	ASTM D 974:2002	0.76	Nil	NA	NA	Nil
3	API gravity @ 60°F	ASTM D 1298:1999	46.67	NA	60.65	NA	31.4
4	Flash point COC	ASTM D 92-05a	<40°C	45°C	22	Plus 1°C	NA
5	Kinematic viscosity @ 40°C	ASTM D 445:2005	2.149 mm^2/s	1.47 mm^2/s	NA	1.63 mm^2/s	5.69
6	Colour	ASTM D 1500:2004a	D 8 (light yellowish-red clear)	NA	Pale yellow	Dark brownish clear	Orange
7	Conradson carbon residue	ASTM D 189:2005	0.010% (wt%)	NA	NA	0.01%	NA
8	Asphaltene content	ASTM D 3279:2001	0.21 (wt%)	NA	NA	NA	NA
9	Ash content	ASTM D 482:2003	<0.01% (wt)	0.02%	NA	NA	NA
10	Calculated carbon aromatic index	ISO 8217:1996	763.4	NA	NA	NA	NA
11	Pour point	ASTM D 97-05a	Minus 10°C	NA	<−20°C	Minus 15°C	1.6°C
12	Sediment by extraction	ASTM D 473:2002	0.012 (wt)	0.001	NA	NA	NA
13	Specific gravity @ 15°C	ASTM D 4052:2002	0.7932	NA	0.7365	0.7835	0.8690
14	Sulphur content	ASTM D 4094:2003	0.0025%	0.083%	0.002%	0.019%	0.09%
15	Water by distillation	ASTM D 95-05el	<0.05%	0.01%	NA	NA	0.05%
16	Calorific value	ASTM D 240 cal/g	10,000	10,810	10,498	10,244	10,460

TABLE 1: *Cont.*

SR. no.	Test	Method of testing	Result	Waste LDPE [20]	Waste mixed plastic [12]	Waste HDPE [27]	Gulf fuel oil [28]
17	Distillation range	ASTM D 86:04b					
	Initial boiling range		71°C			82	
	5% recovery		110°C				
	10% recovery		141°C			126	
	20%		185°C				
	30%		226°C			188	
	40%		261°C				
	50%		291°C			226	
	60%		319°C				
	70%		343°C			278	
	80%		365°C				
	85%		380°C				
	90%		390°C			320	
	100%		400°C			352	
	Total recovery		90%	85.3%		95%	99.6%

NA: not available.

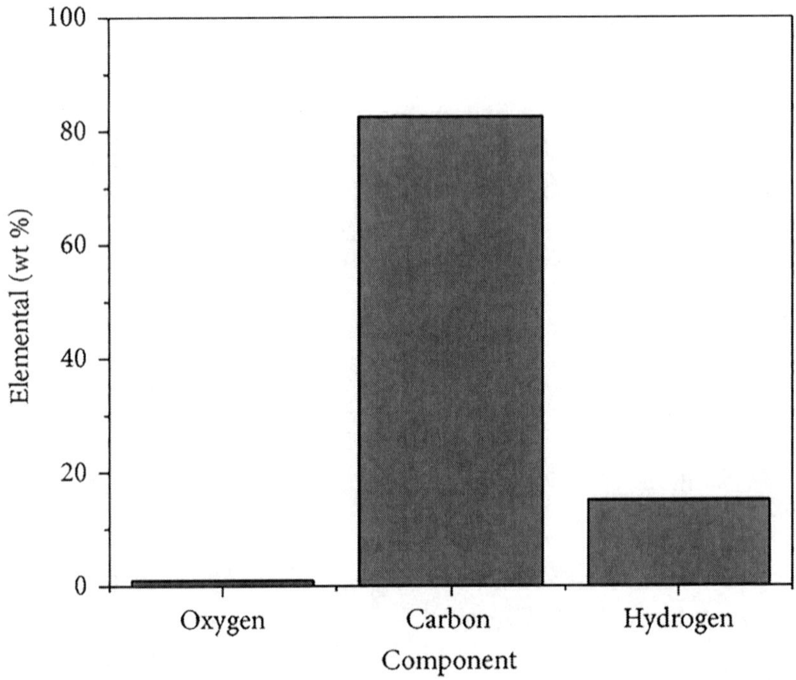

FIGURE 3: Elemental analysis of fuel oil.

Most preliminary analysis of the oil was its visual inspection of the colour obtained after catalytic pyrolysis of WPP. The appearance of oil was transparent and yellowish red in colour. The oil obtained from the pyrolysis was fractionated by distillation and the fuel properties were studied. Elemental composition of fuel oil is found to be C (83%), H (14%), and O (1%) which confirmed hydrocarbon nature of oil (Figure 3). A comparative analysis of sulfur content, pour point, viscosity, and distillation recovery of oil synthesized using various methods is presented in Table 1. Pour point of the sample was found to be less than 10°C, which is in accordance with the standard values of fuel oil [26]. Flash points, density, acidity of fuel oil were found to be ~40°C, 0.7930 g/mL, and 0.76, respectively. All the above values were found to be standard values of fuel oil [11, 27].

American Petroleum Institute (API) gravity of liquefied hydrocarbon prepared from waste plastics was found to be 46.67. There is an inverse relationship between API gravity and density; that is, the higher the density, the lower the API gravity. Light crude is generally that with API gravity over 40 [20]. Therefore, liquefied hydrocarbon produced from our pyrolysis system was found to be light.

Specific gravity of liquefied hydrocarbon prepared from pyrolysis of waste PP was estimated to be 0.7932 at 15°C which is similar to specific gravity of petroleum (density compared to water), that is, 0.8 [20, 29].

Liquefied hydrocarbon synthesized from pyrolysis of WPP has 25 ppm (0.0025%) of sulfur content which is very less compared to standard fuel oil. Fuel oil is normally described as sweet (low sulfur) or sour (high sulfur) depending on their sulfur content. Sweet fuel oil has sulfur content less than 0.5%, and anything more than that is sour. Sweet oil is more preferable than the sour due to its applicability to produce most of the refined products [18, 27]. Another important attribute was found to be kinematic viscosity of liquefied hydrocarbon synthesized from waste plastic that was estimated to be 2.149 mm^2/s at 40°C. As per the standard kinematic viscosities of fuel oil (3.5–9.7 mm^2/s), oil formed by our method was found to be less viscous [20, 30, 31].

Calorific value of oil synthesized from pyrolysis of waste plastic was recorded to be 10,000 cal/g (41.86 MJ/kg), whereas Calorific values of various fuel oils, for example, diesel, gasoline, petrol and petroleum fuel are 44.8, 47.3, 48 and 43 MJ/Kg respectively [20, 30, 31]. Hence, it can be said that this liquefied hydrocarbon obtained from waste plastic can be used for fuel.

From the distillation report of the liquefied hydrocarbon, it was found that boiling range of the oil is 71–390°C, which suggests the presence of a mixture of different fractions such as gasoline, kerosene, and diesel [17, 20, 30–32].

A typical GC chromatogram diagram for the liquid fraction taken from plastics pyrolysis is illustrated in Figure 4. It is clear that chromatogram obtained in high efficient column contains two dominant products which are alk-1-enes and n-alkanes in the range C6–C23, with only a small proportion of cyclic and alcoholic substances. Besides these main mixed plastics decomposition products, chromatogram is fully occupied by several hundred small peaks of other compounds including broad peak of many unresolved compounds which is characteristic for multicomponent hydrocarbon mixtures. This is due to low pyrolysis temperature. Furthermore, a detailed list of all hydrocarbons detected in the liquid fraction of pyrolysis of waste products based on high-density polyethylene (HDPE) and low-density polyethylene (LDPE) was well explained by Achilias et al. in 2007 [8]. It was observed that pyrolysis of the plastic bag made from PE and PP leads to a fraction mainly in the region of C7–C12, which is in the gasoline region. Also, the main components were alkanes and alkenes. A mass spectrum indicates that the fragment of polymers contains only one carbon atom; this is due to the fragmentation at interval of $-CH_2$ break from long chain. Mass spectra at different time intervals show that oil contained alkanes, alkenes, some cycloalkanes, and alcohol. Mass spectra at 3.9 min show that oil might contain cycloalkane and alkenes like cyclodecane and 1-decene, respectively (Figure 5). There are some aliphatic alkanes found to be present at time interval of 4 min having m/z ratio of 142 corresponds to decane ($C_{10}H_{22}$) (Figure 6). It is seen from Figure 7 that there are some higher alcoholic groups like n-pentadecanol ($C_{15}H_{32}O$). Figure 8 indicates higher alkanes, present at interval time of 29.4 minutes; they are nonacosane ($C_{29}H_{60}$) and triacontane ($C_{30}H_{62}$). It is concluded from the GC-MS data that oil contain alkanes, alkenes, cyclic alkanes, and some alcoholic groups.

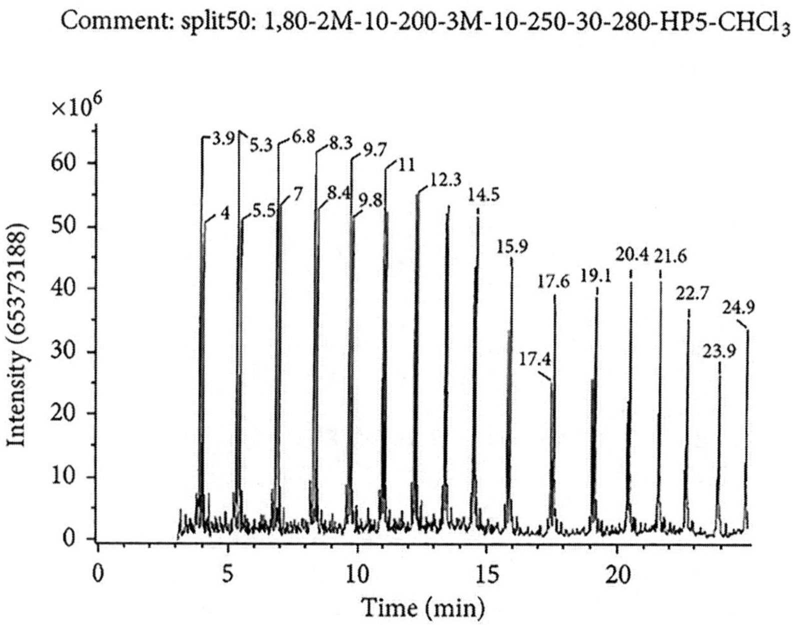

FIGURE 4: Total ion chromatogram of liquefied hydrocarbon obtained from the pyrolysis of waste PP at 550°C in the presence of Ni as catalyst for a duration of 1 hr in atmosphere of Ar.

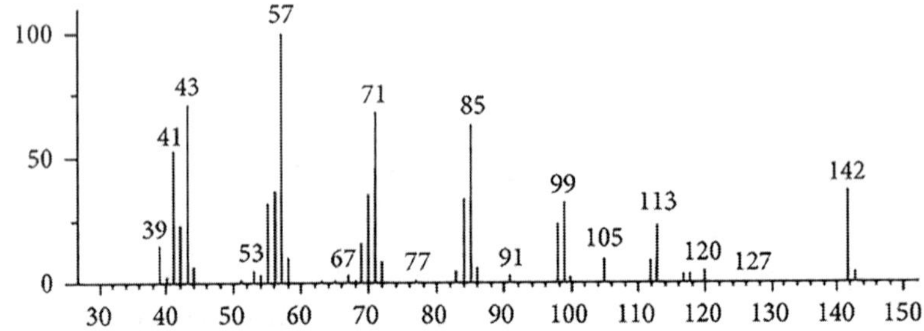

Hit 1: decane
C10H22; MF: 863; RMF: 909; Prob 58.9%; CAS: 124-18-5; Lib: mainlib; ID: 21679

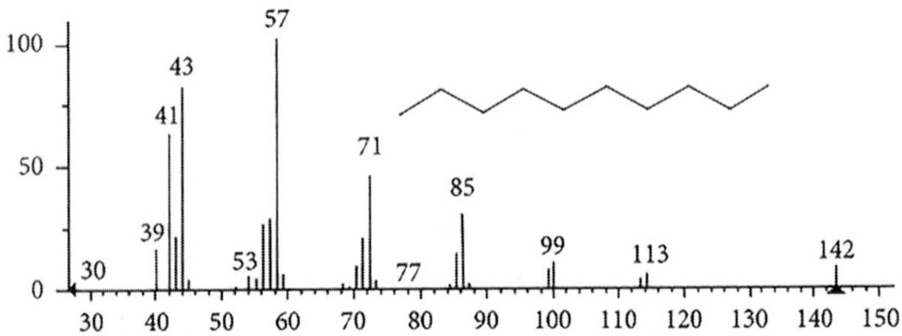

Hit 2: decane
C10H22; MF: 832; RMF: 880; Prob 58.9%; CAS: 124-18-5; Lib: replib; ID: 2052

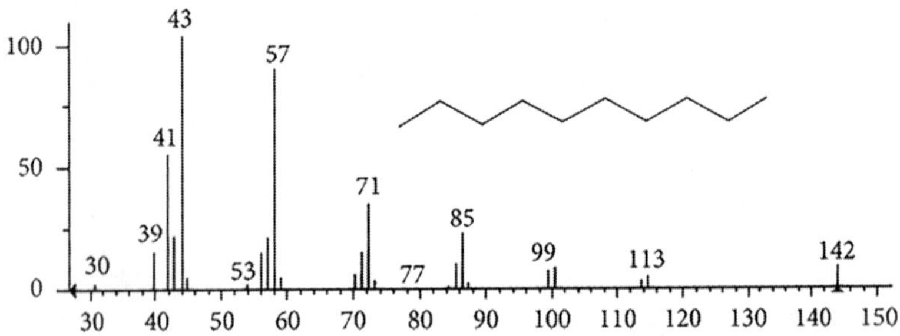

FIGURE 5: Mass spectra of peaks at 3.9 min.

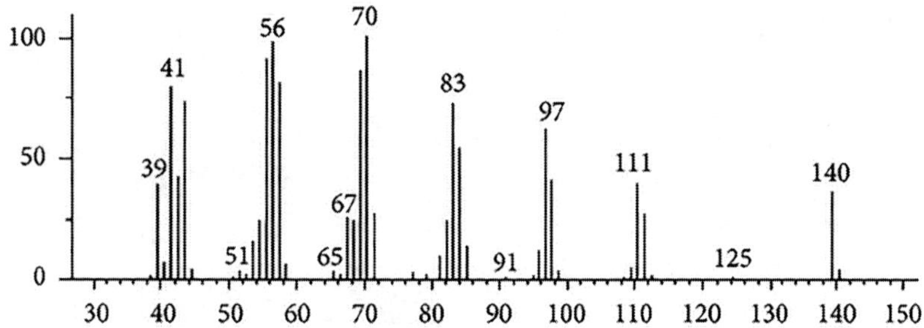

Hit 1: cyclodecane
C10H20; MF: 906; RMF: 906; Prob 13.4%; CAS: 293-96-9; Lib: replib; ID: 4285

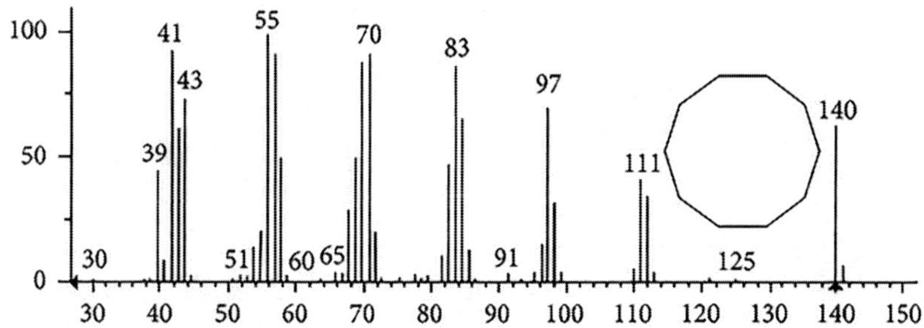

Hit 2: 1-decane
C10H20; MF: 901; RMF: 904; Prob 10.8%; CAS: 872-05-9; Lib: replib; ID: 4200

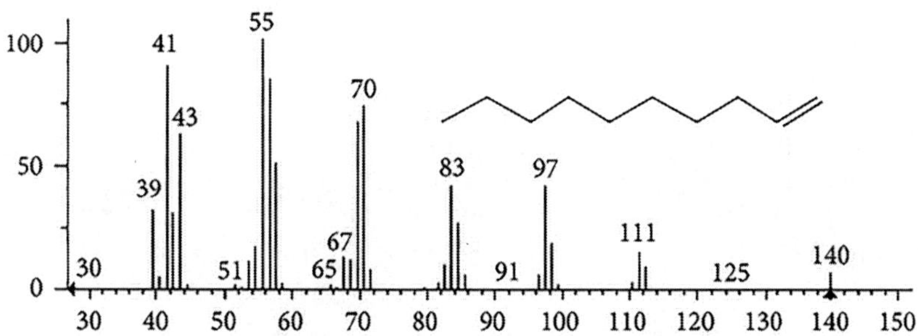

FIGURE 6: Mass spectra of peaks at 4 min.

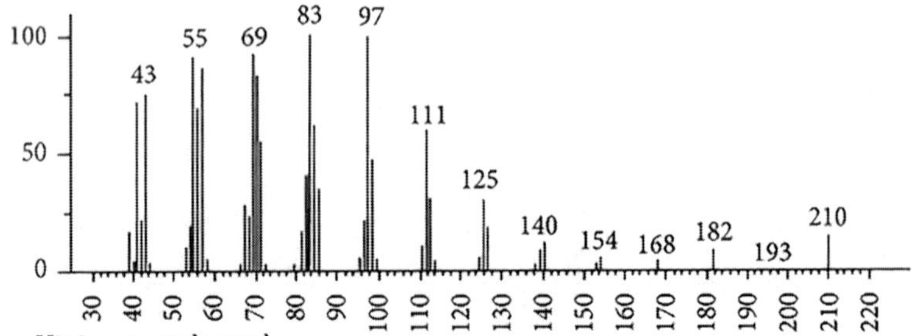

Hit 1: n-pentadecanol
C15H32O; MF: 914; RMF: 915; Prob 12.1%; CAS: 629-76-5; Lib: replib; ID: 1980

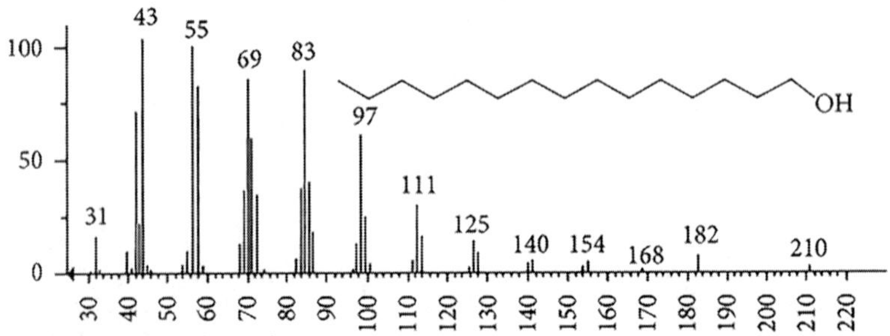

Hit 2: n-pentadecanol
C15H32O; MF: 909; RMF: 912; Prob 12.1%; CAS: 629-76-5; Lib: mainlib; ID: 44698

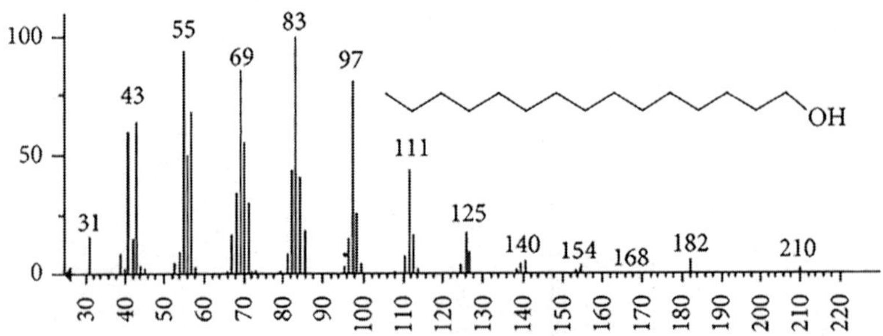

FIGURE 7: Mass spectra of peaks at 11 min.

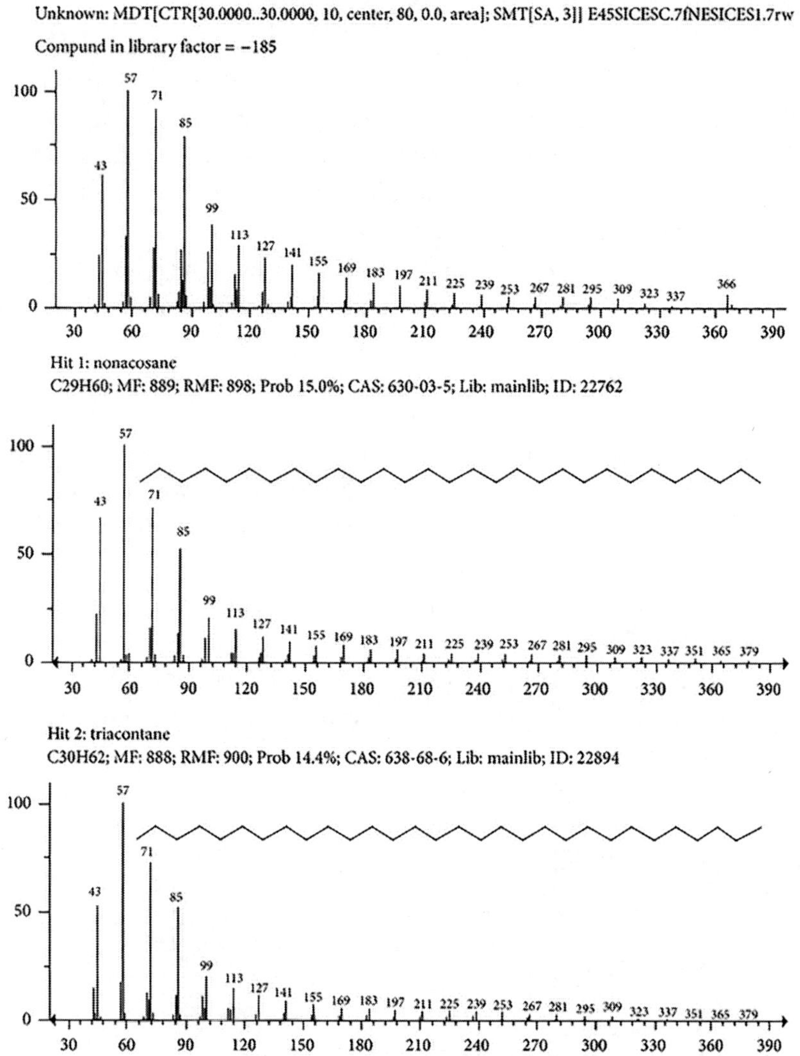

FIGURE 8: Mass spectra of peaks at 29.4 min.

9.5 MECHANISM OF CATALYTIC DEGRADATION OF PLASTICS

The presence of a small amount of Ni catalyst could efficiently promote the dehydrogenation of PP into light hydrocarbons. It was because that the Ni catalyst attacked PP surface to form cationic active sites and promoted the formation of more molecules with a lower carbon number by a cationic mechanism [14, 15]. Catalysts such as metal oxides have appeared to be used mainly for enhancement of monomer recovery [32]. Degradation of PP on nickel oxide solution yield more oils than that on solid acids, and time required to complete degradation on nickel oxide is lower than on solid acids. The composition of oil on metal oxide (NiO) is reported to be rich in 1-olefins and is poor in aromatics and branched isomers. In our case, plastics reacted with catalyst and were converted into smaller fragments or monomers. As the temperature increases these fragments are converted into vapour state which condense to give fuel oil. Octane number is expected to be low for the oils produced on solid bases, since the oils mainly consisted of straight chain hydrocarbons: n-paraffins, and 1-olefins. From an economic point of view, reducing the cost even further will make this process an even more attractive option. This option can be optimized by reuse of catalysts and the use of effective catalysts in smaller quantities [28, 33]. This method seems to be the most promising to be developed into a cost-effective commercial plastic waste recycling process to solve the keen environmental problem of plastic waste disposal.

9.6 CONCLUSION

Catalytic degradation of waste plastics in the presence of Ni catalyst for duration of 1 h in presence of Ar atmosphere leads to following concluding remarks.

1. Conversion efficiency of Ni catalysed pyrolysis of WPP was found to be more than 90% yield. The catalytic degradation pro-

cess produces much less residue content than that from thermal degradation.
2. Physicochemical properties of obtained fuel oil can be exploited to make highly efficient fuel or furnace oil after blending with other petroleum products.
3. The knowledge of design and process of the semiscale plant will be helpful for developing a commercial scale plant in the future. From this result, it can be concluded that the fuel properties of the catalytic pyrolysis oil match the properties of petroleum fuels.

HIGHLIGHTS

1. Ecofriendly method for waste management of polypropylene plastics.
2. Use of nano-sized catalyst to enhance the rate of conversion of waste plastics to fuel oil.
3. Ideal properties of fuel oil for efficient commercial usage.
4. Unique design of the furnace to facilitate efficient conversion.

REFERENCES

1. G. Luo, T. Suto, S. Yasu, and K. Kato, "Catalytic degradation of high density polyethylene and polypropylene into liquid fuel in a powder-particle fluidized bed," Polymer Degradation and Stability, vol. 70, no. 1, pp. 97–102, 2000.
2. N. Miskolczi, L. Bartha, and A. Angyal, "High energy containing fractions from plastic wastes by their chemical recycling," Macromolecular Symposia, vol. 245-246, pp. 599–606, 2006.
3. A. Marcilla, M. I. Beltrán, and R. Navarro, "Evolution of products generated during the dynamic pyrolysis of LDPE and HDPE over HZSM5," Energy and Fuels, vol. 22, no. 5, pp. 2917–2924, 2008.
4. Y. Ando, X. Zhao, T. Sugai, and M. Kumar, "Growing carbon nanotubes," Materials Today, vol. 7, no. 9, pp. 22–29, 2004.
5. D. Pradhan, M. Sharon, M. Kumar, and Y. Ando, "Nano-octopus: a new form of branching carbon nanofiber," Journal of Nanoscience and Nanotechnology, vol. 3, no. 3, pp. 215–217, 2003.

6. S. Iijima, "Helical microtubules of graphitic carbon," Nature, vol. 354, no. 6348, pp. 56–58, 1991.
7. P. S. Umare, R. Antony, K. Gopalakrishnan, G. L. Tembe, and B. Trivedi, "Synthesis of low molecular weight polyethylene waxes by a titanium BINOLate-ethylaluminum sesquichloride catalyst system," Journal of Molecular Catalysis A, vol. 242, no. 1-2, pp. 141–150, 2005.
8. D. S. Achilias, C. Roupakias, P. Megalokonomos, A. A. Lappas, and V. Antonakou, "Chemical recycling of plastic wastes made from polyethylene (LDPE and HDPE) and polypropylene (PP)," Journal of Hazardous Materials, vol. 149, no. 3, pp. 536–542, 2007.
9. R. Aguado, M. Olazar, M. J. San José, B. Gaisán, and J. Bilbao, "Wax formation in the pyrolysis of polyolefins in a conical spouted bed reactor," Energy and Fuels, vol. 16, no. 6, pp. 1429–1437, 2002.
10. W. Kaminsky, M. Predel, and A. Sadiki, "Feedstock recycling of polymers by pyrolysis in a fluidised bed," Polymer Degradation and Stability, vol. 85, no. 3, pp. 1045–1050, 2004.
11. M. Sarker, M. M. Rashid, and M. Molla, "Abundant High-Density Polyethylene (HDPE-2) turns into fuel by using of HZSM-5 catalyst," Journal of Fundamentals of Renewable Energy and Applications, vol. 1, pp. 1–12, 2012.
12. D. C. Tiwari, E. Ahmad, and K. Singh, "Catalytic degradation of waste plastic into fuel range hydrocarbons," International Journal of Chemical Research, vol. 1, pp. 31–36, 2009.
13. Y.-H. Lin and H.-Y. Yen, "Fluidised bed pyrolysis of polypropylene over cracking catalysts for producing hydrocarbons," Polymer Degradation and Stability, vol. 89, no. 1, pp. 101–108, 2005.
14. L. Soják, R. Kubinec, H. Jurdáková, E. Hájeková, and M. Bajus, "High resolution gas chromatographic-mass spectrometric analysis of polyethylene and polypropylene thermal cracking products," Journal of Analytical and Applied Pyrolysis, vol. 78, no. 2, pp. 387–399, 2007.
15. N. Mishra, G. Das, A. Ansaldo et al., "Pyrolysis of waste polypropylene for the synthesis of carbon nanotubes," Journal of Analytical and Applied Pyrolysis, vol. 94, pp. 91–98, 2012.
16. F. Pinto, P. Costa, I. Gulyurtlu, and I. Cabrita, "Pyrolysis of plastic wastes—1. Effect of plastic waste composition on product yield," Journal of Analytical and Applied Pyrolysis, vol. 51, no. 1, pp. 39–55, 1999.
17. B. Tymoshevskyy, L. Malyi, M. Tkach, and G. Bykovchenko, "Utilization efficiency of aramide polymeric materials of the cases of solid-propellant missiles," World Academy of Science, Engineering and Technology, vol. 52, pp. 80–82, 2009.
18. J.-G. Na, B.-H. Jeong, S. H. Chung, and S. Kim, "Pyrolysis of low-density polyethylene using synthetic catalysts produced from fly ash," Journal of Material Cycles and Waste Management, vol. 8, pp. 126–132, 2006.
19. A. R. Songip, T. Masuda, H. Kuwahara, and K. Hashimoto, "Test to screen catalysts for reforming heavy oil from waste plastics," Applied Catalysis B, vol. 2, no. 2-3, pp. 153–164, 1993.

20. S. Pandian and A. Kamalakannan, "Catalytic pyrolysis of dairy industrial waste LDPE film into fuel," International Journal of Chemical Research, vol. 3, pp. 52–55, 2012.
21. K. N. Kim and S.-G. Kim, "Nickel particles prepared from nickel nitrate with and without urea by spray pyrolysis," Powder Technology, vol. 145, no. 3, pp. 155–162, 2004.
22. A. R. Ardiyanti, S. A. Khromova, R. H. Venderbosch, V. A. Yakovlev, and H. J. Heeres, "Catalytic hydrotreatment of fast-pyrolysis oil using non-sulfided bimetallic Ni-Cu catalysts on a δ-Al2O3 support," Applied Catalysis B, vol. 117-118, pp. 105–117, 2012.
23. X. Zhang, T. Wang, L. Ma, Q. Zhang, and T. Jiang, "Hydro treatment of bio-oil over Ni-based catalyst," Bioresource Technology, vol. 127, pp. 306–311, 2012.
24. K.-H. Lee and D.-H. Shin, "Catalytic degradation of waste polyolefinic polymers using spent FCC catalyst with various experimental variables," Korean Journal of Chemical Engineering, vol. 20, no. 1, pp. 89–92, 2003.
25. Z. Jiang, R. Song, W. Bi, J. Lu, and T. Tang, "Polypropylene as a carbon source for the synthesis of multi-walled carbon nanotubes via catalytic combustion," Carbon, vol. 45, no. 2, pp. 449–458, 2007.
26. N. X. Thanh, M. Hsieh, and R. P. Philp, "Waxes and asphaltenes in crude oils," Organic Geochemistry, vol. 30, no. 2-3, pp. 119–132, 1999.
27. S. Kumar and R. K. Singh, "Recovery of hydrocarbon liquid from waste high density polyethylene by thermal pyrolysis," Brazilian Journal of Chemical Engineering, vol. 28, no. 4, pp. 659–667, 2011.
28. G. Chauhan, K. K. Pant, and K. D. P. Nigam, "Extraction of nickel from spent catalyst using biodegradable chelating agent EDDS," Industrial & Engineering Chemistry Research, vol. 51, pp. 10354–10363, 2012.
29. P. Hudec, M. Horňáček, A. Smiešková, and P. Duačík, "Chemical recycling of waste hydrocarbons in catalytic cracking," Petroleum & Coal, vol. 51, pp. 51–58, 2009.
30. S. Murugan, M. C. Ramaswamy, and G. Nagarajan, "The use of tyre pyrolysis oil in diesel engines," Waste Management, vol. 28, no. 12, pp. 2743–2749, 2008.
31. M. Sarker, M. M. Rashid, M. S. Rahman, and M. Molla, "Alternative diesel grade fuel transformed from polypropylene (PP) municipal waste plastic using thermal cracking with fractional column distillation," Energy and Power Engineering, vol. 4, pp. 165–172, 2012.
32. A. G. Buekens and H. Huang, "Catalytic plastics cracking for recovery of gasoline-range hydrocarbons from municipal plastic wastes," Resources, Conservation and Recycling, vol. 23, no. 3, pp. 163–181, 1998.
33. S. Goel, K. K. Pant, and K. D. P. Nigam, "Extraction of nickel from spent catalyst using fresh and recovered EDTA," Journal of Hazardous Materials, vol. 171, no. 1-3, pp. 253–261, 2009.

CHAPTER 10

Kinetic Study of the Pyrolysis of Waste Printed Circuit Board Subject to Conventional and Microwave Heating

JING SUN, WENLONG WANG, ZHEN LIU, QINGLUAN MA, CHAO ZHAO, AND CHUNYUAN MA

NOMENCLATURE

- A: Pre-exponential factor, s^{-1}
- E: Activation energy, kJ/mol
- f: Function of conversion
- k: Thermal decomposition rate constant, s^{-1}
- n: Reaction order
- R: Gas constant = 8.314 kJ/(kmol K)
- t: Pyrolysis time, s
- T: Absolute temperature, K
- T_0: The temperature at which the pyrolysis of WPCB begin, °C
- T_{1m}: The temperature at which the DTG curve of WPCB reaches the first peak, °C

- T_{2m}: The temperature at which the DTG curve of WPCB reaches the second peak, °C
- W: Weight of sample at time t, mg and/or g
- W_0: Initial weight of sample, mg and/or g
- W_∞: Final weight of sample, mg and/or g
- x: Conversion of WPCB, defined as: $x = w_0 - w/(w_0 - w_\infty)$
- Greek letters: Descriptions
- β: Heating rate, °C min^{-1}

10.1 INTRODUCTION

Due to the rapid development of technology, the lifespan of electrical and electronic equipment (EEE) is very short, resulting in increasing quantities of waste electronic and electrical equipment (WEEE). It is estimated that some 40–60 million tons of WEEE are generated worldwide every year, which poses grave risks to human health and the environment [1,2]. However, in spite of its hazardous properties, WEEE is also considered a precious resource, so the development of effective WEEE recycling techniques is of increasing interest. As one of the most important branches of the WEEE stream, waste printed circuit boards (WPCB) are generally considered to be representative of WEEE, and have received increasing attention from the public and researchers because they contain a wealth of nonferrous metals in addition to a variety of toxic materials [3,4]. Thus, the recycling of WPCB is important not only as regards environmental protection and the recovery of valuable materials, but also for providing guidance to WEEE recycling.

Pyrolysis can be described as the thermal decomposition of organic components in an oxygen-free atmosphere to yield char, oil, and gas, and it holds promise as an approach to optimal waste upgrading, especially for organic polymers. Moreover, operational conditions in pyrolytic processes can be optimized to improve the yield or the quality of desirable products by means of innovative heating methods or catalysts. Microwave pyrolysis (MWP) is one focus of current research and has drawn the widespread interest of foreign and domestic researchers [5–8]. The applications of MWP include the pyrolysis of plastic waste [5,6], the pyrolysis of biomass waste [7–13], sewage sludge [14,15], and waste automotive engine oil [16–18], among others.

Although several groups have studied the pyrolysis of WPCB, their focus has mainly been on the pyrolytic kinetics [19–22] and the formation and fate of brominated compound [22–24] in conventional heating schemes (such as electric heating), also known as conventional pyrolysis. Due to the presence of metals, which are considered incompatible with microwave heating, the microwave-induced pyrolysis of WPCB has barely been studied. Under microwave irradiation, a discharge will be triggered when there are metal tips or sharp corners in the material being treated. This is a result of field enhancement and field emission of electrons when the induced electric potential exceeds the coulomb potential. As a result of these discharges, metal tips can be melted, implying that a discharge can produce local temperatures of 1000 °C–2000 °C [25–27]. Our research group has studied the heating effects associated with microwave-metal discharges and found that the energy conversion ratio from electrical energy to heat can reach as high as 30% [28]. Based on the unique material characteristics of WPCBs and the discharge properties of metals, we have previously established that microwave-induced pyrolysis holds promise as a way to process WPCBs [26–30].

Kinetics studies of microwave-induced pyrolysis of WPCBs have rarely been performed. However, an understanding of the pyrolytic kinetics is important for reactor design, process optimization, and the general improvement of microwave heating applications. This calls for the development of thermal decomposition kinetic models that can accurately predict the pyrolysis processes that occur in WPCBs under microwave irradiation. The prevailing thermal gravimetric analysis (TGA) technique is able to predict such thermal decomposition processes quite accurately, and is considered to be representative of conventional pyrolysis modeling. In this work, the pyrolytic kinetics of WPCBs subject to both conventional heating and microwave heating are studied. We present and compare kinetic parameters of interest that include the activation energy, a pre-exponential factor, and the reaction order in the chemical reaction-controlled region. These data may be of use in recycling process design and in energy recovery schemes for WPCB reuse in pyrolytic processes.

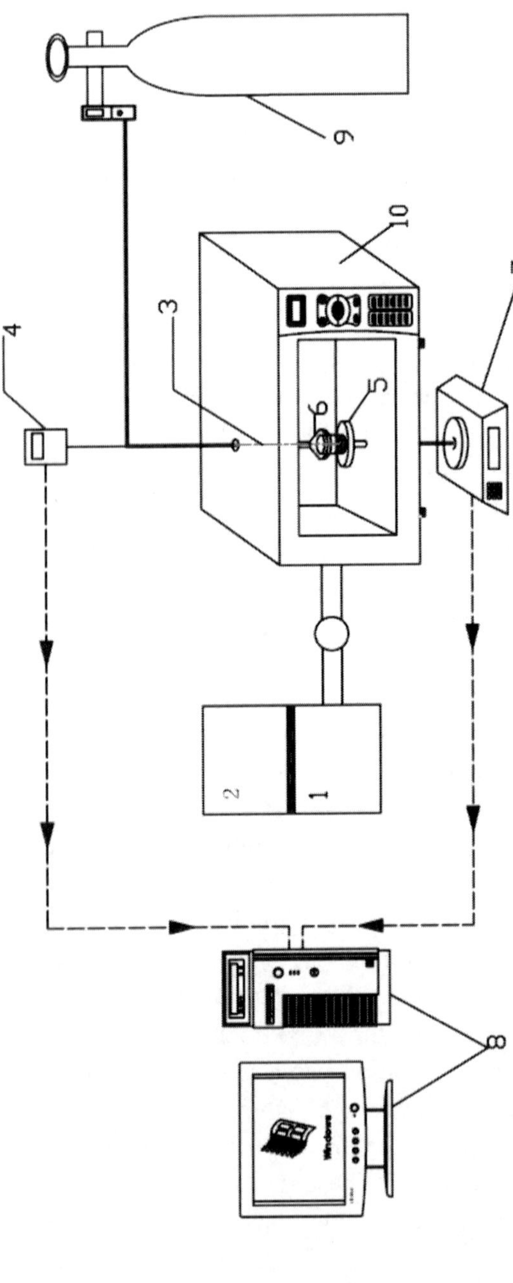

Figure 1. Experimental set-up for microwave pyrolysis of WPCB: (1) microwave generation system; (2) microwave control system; (3) thermocouple; (4) temperature monitoring system; (5) PTFE tray system; (6) quartz container; (7) weighing system; (8) signal acquisition and processing system; (9) nitrogen and (10) microwave oven.

10.2 MATERIALS AND METHODS

10.2.1 MATERIALS AND SAMPLE PREPARATION

Waste samples of printed circuit boards, obtained from a recycling plant in Jinan, China, served as the raw material for our studies. The WP-CBs were first shredded into fragments of about 1 cm × 1 cm (referred to as "big particles"), and then a representative fragment was sampled and crushed into a fine particle state (approximately 40–60 mesh) in the laboratory. The proximate analysis of WPCBs was carried out by referring to the coal proximate analysis method. The weight percentages of moisture, ash, volatiles and fixed carbon were 0.48%, 73.62%, 25.12 % and 0.78%, respectively.

10.2.2 APPARATUS AND PROCEDURES

The pyrolysis experiments were carried out in a thermogravimetric analyzer (Mettler-Toledo, TGA/DSC/1600HT) for the conventional heating data, and in a modified household microwave oven (Galanz P7021TP-6) for the microwave-induced pyrolysis experiments. For the conventional TGA of WPCB, about 15.0 mg of fine particle material was placed in a crucible pan and heated in an inert atmosphere of nitrogen (N_2 flow rate of 20 mL/min) over a temperature range of 50–900 °C at heating rates of 10, 20, 30 and 50 °C/min. The temperature and weight of the material was monitored with a computer. For the microwave-induced pyrolysis of WPCB, 40 g of "big particle" material was placed in a quartz tube (60 mm length, 55 mm i.d.) which was itself surrounded by a quartz container and purged with nitrogen (99.999% pure) at a flow rate of 100 mL/min. During microwave irradiation, the temperature was measured by two ways: one is an online measurement by placing the thermocouple 5 mm directly above the sample to monitoring the temperature of volatility; the other is an offline measurement by setting different irradiation time (e.g., 40 s, 60 s, 90 s, 120 s, 150 s, 3 min, 5 min,…) and inserting the thermocouple into the center of the sample immediately after turning off microwaves. The thermocouple we used was connected with a filter capacity to reduce the disturbance from microwaves and improve the reliability of the temperature measurement. The offline temperature measurement was used to correct the online measurement results by averaging at certain time points. Concurrently, the sample mass when the thermocouple centered 5 mm directly above was recorded via an electronic balance system underneath. Figure 1 shows a diagram of the experimental setup.

10.2.3 PYROLYSIS KINETICS

Pyrolysis of WPCB is generally a complex process. It is difficult to discover a full kinetic analysis of complex systems, but some kind of "effective" or "average" kinetic description is still needed. In this work, the overall rate of the pyrolytic reactions of WPCB is described by the following equation:

$$dx/dt = kf(x) \tag{1}$$

where f(x) is a function depending on the decomposition mechanism, which can be in the form of an nth order of reaction: $f(x) = (1 - x)^n$.

Constant k obeys the Arrhenius correlation:

$$k = A\exp(-E/RT) \tag{2}$$

Rearranging and integrating, we have:

$$\int_0^x \frac{dx}{(1-x)^n} = \frac{Z}{\beta}\int_{T_0}^{T} \exp(-E/RT)\,dT \tag{3}$$

Based on Equation (3), Coats-Redfern's method was derived for numerical determination of the kinetic parameters, as Equations (4) and (5):

when n ≠ 1:

$$\ln\left\{\frac{1-(1-x)^{1-n}}{T^2(1-n)}\right\} = \ln\left[\frac{AR}{\beta E}\left(1-\frac{2RT}{E}\right)\right] - \frac{E}{RT} \tag{4}$$

when n = 1:

$$\ln\left\{\frac{-\ln(1-x)}{T^2}\right\} = \ln\left[\frac{AR}{\beta E}\left(1-\frac{2RT}{E}\right)\right] - \frac{E}{RT} \tag{5}$$

When 2RT/E is small enough to be ignored, Equations (4) and (5) can be simplified as Equations (6) and (7):

when n ≠ 1:

$$\ln\left\{\frac{1-(1-x)^{1-n}}{T^2(1-n)}\right\} = \ln\left(\frac{AR}{\beta E}\right) - \frac{E}{RT} \tag{6}$$

when n = 1:

$$\ln\left\{\frac{-\ln(1-x)}{T^2}\right\} = \ln\left(\frac{AR}{\beta E}\right) - \frac{E}{RT} \tag{7}$$

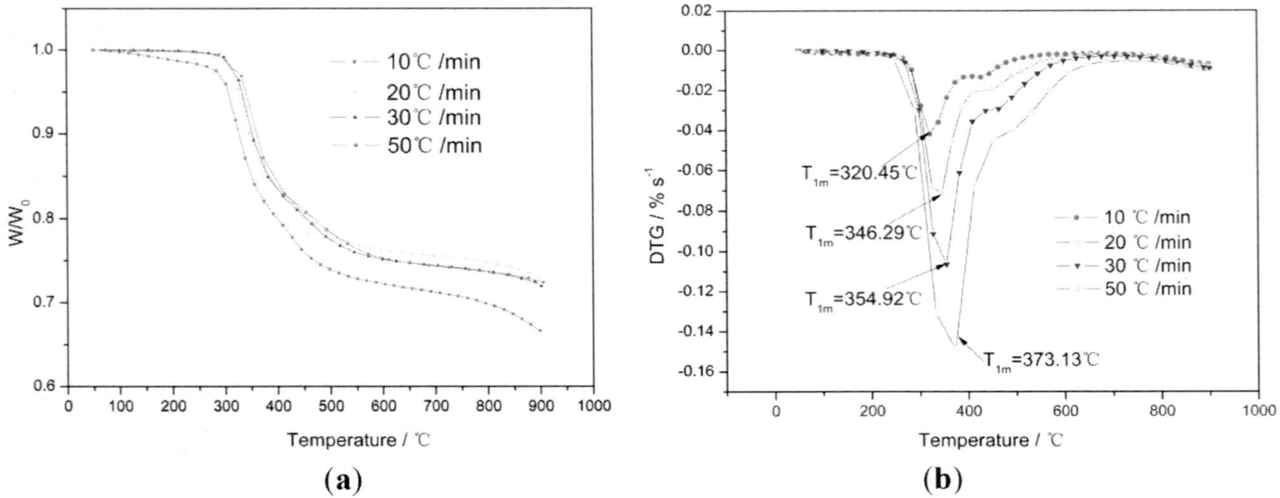

FIGURE 2: Thermogravimetry curves (a) and derivative thermogravimetry curves (b) at different heating rates.

With regards the scenario that RT/E is not negligible, the Equations (4) and (5) can be transformed into Equations (8) and (9):

when $n \neq 1$:

$$\ln\left\{\frac{1-(1-x)^{1-n}}{T^2(1-n)\left(1-\frac{2RT}{E}\right)}\right\} = \ln\frac{AR}{\beta E} - \frac{E}{RT} \quad (8)$$

when $n = 1$:

$$\ln\left\{\frac{-\ln(1-x)}{T^2\left(1-\frac{2RT}{E}\right)}\right\} = \ln\frac{AR}{\beta E} - \frac{E}{RT} \quad (9)$$

Iterative method and the least squares method should be combined to determine the value of n, A and E.

10.3 RESULTS AND DISCUSSION

10.3.1 CONVENTIONAL PYROLYSIS OF WPCBS USING TGA

The TG and DTG curves from the conventional heating experiments are shown in Figure 2. As evidenced by the DTG curve, there may be two overlapped degradation stages due to the presence of a turning point around 430 °C. However, there were no obvious divide between these two stages, thus it can be considered that one degradation stage was found during the pyrolysis process, characterized by an obvious weight loss peak. The main decomposition occurred at 300 °C–600 °C with an accompanying conversion of about 90%. The main thermal degradation temperature in this paper were a little high when compared to the work of Chen et al. [19] and Chiang et al. [31] which could be caused by high heating rates and different compositions. Thus, the general weight loss in the WPCBs can be characterized by the following stages: the weight of the sample barely changed when its temperature remained below 300 °C; thereafter the sample underwent a rapid weight loss in the temperature range of 300 °C–600 °C. This weight loss process comprised the pyrolysis phase, where about 25% of the original weight was lost. Subsequently, the weight of the sample was largely maintained with a very slow weight loss process.

With increased heating rates, the characteristic temperatures of T_0, T_{1m}, T_{2m} (description in nomenclature, T_{2m} was not marked in Figure 2 due to the vague division between the second peak and the first one) were increased, the DTG curve was shifted towards higher temperature regions, and the value of dx/dt and the maximum decomposition rates were also improved. We can conclude that the pyrolysis reaction was postponed to higher temperature range with increasing heating rates, because the time interval in which the sample is exposed to a given temperature decreases as the heating rate increases. This result is in accordance with the work of other researchers. The final residual weight rate was strongly influenced by variations in the heating rate. The residual weight rate decreased as a result of mass-loss during thermal decomposition, indicating volatile matter was ejected. With low heating rates, the time interval in which the sample is exposed to a given temperature is long, so that the sample has ample time to decompose, improving the reaction depth. As many metal components including the metals with low melting or boiling points, such as Hg, Sn, Pb, Zn, etc., were contained in WPCBs, long residence time may intensified the volatility or emission of heavy metal components. Moreover, compared to the proximate analysis results of WPCBs, WPCBs can obtain a complete pyrolysis with the heating rates from 20 °C/min to 50 °C/min. Thus, the following kinetic study of the pyrolysis process is based mainly on the heating rates of 20, 30 and 50 °C/min.

As most materials decompose (about 90% conversion occurred) in the temperature range of 300 °C–600 °C, which comprises the main reaction region, the kinetic study of the thermal decomposition in this temperature range is the main focus of our work. The Coats-Redfern method was used to determine the kinetic parameters for heating rates of 20, 30, and 50 °C/min and their values are shown in Table 1.

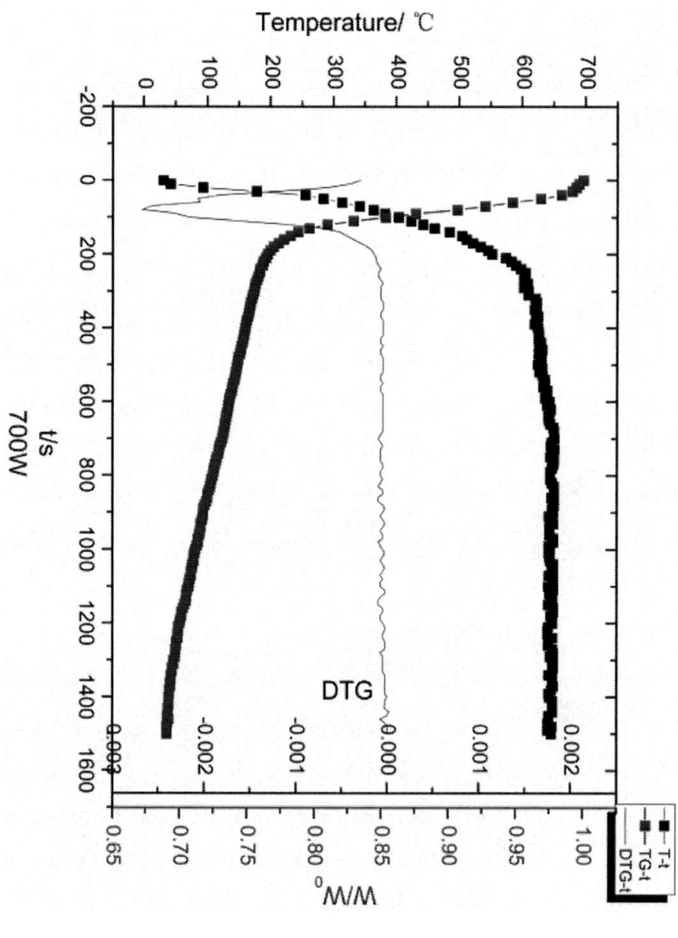

FIGURE 3: T-t, TG-t, and DTG-t curves of WPCBs in the microwave pyrolysis process.

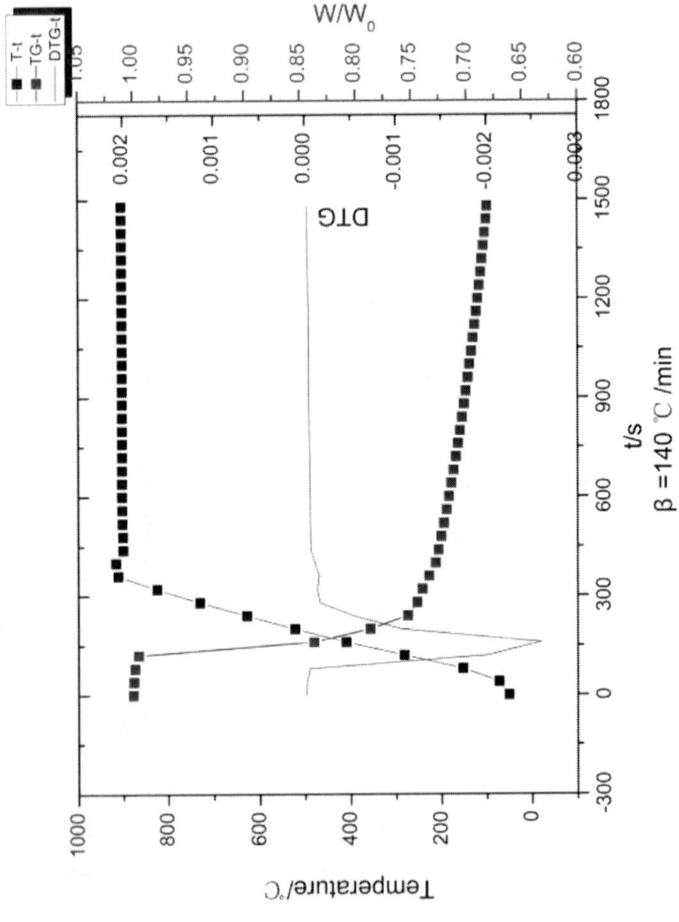

FIGURE 4: T-t, TG-t, and DTG-t curves of WPCBs in simulation of the conventional heating process with a heating rate of 140 °C/min.

TABLE 1: Kinetic parameters for conventional pyrolysis of WPCB at three heating rates.

β(°C/min)	n	E	lnA	R^2
20	8.97	266.52	53.51	0.996
30	7.36	219.62	43.69	0.997
50	5.46	168.78	33.22	0.997

The activation energy was found to be 168–267 kJ/mol, decreasing with an increase in heating rate. Our results indicate a significant difference in activation energy for different heating rates. The variation of the activation energy with heating rate gives rise to a variation in the pre-exponential factor. This can be explained with a kinetic compensation effect that can be expressed as a linear relationship between the logarithm of the pre-exponential factor and the activation energy ($\ln(A) = k_E + b$). This effect is mainly due to mathematical, physio-chemical, and experimental causes, which is widely researched [32,33]. Although this coupling of the kinetic parameters can result in similar values of the kinetic constant, the lower activation energy indicates the energy required to activate molecules to start the chemical reaction is lower because chemical reaction occurs on the condition that the molecular energy is equal to or higher than the activation energy. And more molecules will be activated at a given temperature, if the activation energy is lower. Although it is hard to observe the differences of the start of decomposition at different heating rates in Figure 2, which are mainly due to reaction shifts to a high temperature range with the increased heating rate, the reaction rate at a given temperature was obviously improved with the increase in heating rate, as illustrated in Figure 2b. The accelerated reaction rate can be attributable to the strengthened activation of molecules due to lower activation energy. The reaction order was also decreased with an increase in heating rate which demonstrated the thermal decomposition process was very complex when the heating rate was low. And a high heating is beneficial to eliminate unnecessary reactions to make the decomposition process simple.

10.3.2 MICROWAVE-INDUCED PYROLYSIS OF WPCB

The heating rate of WPCB material subject to microwave irradiation is determined by many factors, such as the sample's wave-absorption capacity and the heating effect from microwave-metal discharges. Thus, in contrast to the heating rates in conventional heating schemes, it is much easier to control the microwave power in experiments involving microwave-induced pyrolysis of WPCBs. Figure 3 presents the temperature, residual weight rate and DTG of WPCBs as a function of time when exposed to an incident microwave power of 700 W at 2.45 GHz. The temperature of the sample clearly skyrockets at the beginning of these experiments, and then is followed by an approximately linear increase with a measured heating rate of about 140 °C /min in the range of 250 °C–500 °C, followed by an increase in temperature to above 600 °C at a lower rate of about 60 °C / min, and finally it stabilizes at a temperature of about 650 °C. Correspondingly, the decomposition of WPCBs can be divided into two stages: the sample was rapidly pyrolyzed with a rapid rate of weight loss at temperatures below 500 °C, after which the samples experienced a slow weight loss process.

For a better comparison, a simulation of microwave-induced pyrolysis of WPCB material was carried out by modeling a conventional TGA with a heating rate of 140 °C/min from 50 °C to 900 °C, and subsequent temperature maintenance at 900 °C for 20 min. In fact, it is hard to simulate the heating process of WPCB under microwave irradiation completely. Thus, a simulation of the main pyrolysis phase was simulated. Figure 4 presents the temperature, residual weight rate and DTG of WPCB as a function of time.

Compared with Figure 3, which shows real data from microwave experiments, the overall profiles are similar, except for the lack of an induction phase in the microwave-induced pyrolysis data, since metal-microwave sparks occurred immediately at the onset of microwave irradiation, leading to high local temperatures and a partial decomposition of WPCBs which results in significant weight loss, even though the average overall temperature may be low. As there is only one obvious peak in the data for the pyrolysis stage where the main reaction happened in both pyrolysis processes, our thermal kinetic study will be focused on this range.

It should be noted that the value of the pre-exponential factor A and the activation energy E is a number that is conditioned upon a constant heating rate during the entire pyrolysis process. For the microwave-induced pyrolysis experiments, the temperature curves have an approximately linear phase in the temperature range of 250 °C–500 °C.

Thus, the heating rate during this phase can be considered constant. The following kinetic parameters were calculated based on the data related to this phase. For the simulated TGA, the main pyrolysis phase starts at 280 °C, and this linear phase lies in the temperature range of 280 °C–530 °C. The values of the calculated kinetics parameters are shown in Table 2.

TABLE 2: Kinetic parameters for pyrolysis of WPCB under both conventional TGA and microwave heating scheme.

Pyrolysis method	n	E	lnA	R^2
Microwave power	2.41	48.68	9.50	0.992
Simulated TGA	6.90	106.20	22.10	0.990

Two main differences can be easily observed between the microwave pyrolysis process and the conventional process: one is the sequence of events at the beginning of the pyrolysis reaction; the other one is the activation energy of the main pyrolysis phase. As to the differences in the pyrolysis start, WPCB was pyrolyzed with weight loss at the onset of microwave irradiation, compared with an induction phase of about 100 s in the simulated pyrolysis process. The promotion effect on reaction start caused by microwave irradiation is mainly due to metal-microwave discharge as mentioned above. With regard to the activation energy, it should be same under the same heating rates if the kinetic processes in both cases are same. The fact that the activation energy in microwave heating is remarkably lower than that in conventional heating indicates that the kinetics in microwave heating is different from that in conventional heating schemes. With respect to the application of microwave heating to chemical reactions, both "thermal effect" and "non-thermal effect" are put forward and discussed. Here, the kinetics for microwave pyrolysis process can be attributable to the microwave internal heating style, wherein samples are heated uniformly, leading to reductions in temperature gradients in the sample and a reduced heat loss via conduction: more heat in the sample facilitates chemical reactions [33]. Moreover, the microwave-metal discharge plasma may be also beneficial in driving chemical reactions, as a result of "non-thermal" effects or a thermal catalyst-type effect. Due to the remarkably reduced activation energy caused by microwave heating style, certain special effects (i.e., non-thermal effects) can be concluded to contribute to the reaction kinetics.

In conclusion, the pyrolysis kinetics of WPCB subject to microwave heating are different from those in conventional heating, which makes the research on microwave-induced pyrolysis of waste very worthwhile. The results will help us further our research on microwave pyrolysis of electronic waste and other similar waste. And, more work is required to investigate the non-thermal effect of microwave heating on activation energy in the future.

10.4 CONCLUSIONS

This work describes a kinetic study of the thermal decomposition of WPCBs subject to both conventional TGA electric heating and a microwave thermogravimetric system, although the experimental conditions such as particle size, sample mass, flow rate of purge gas, etc., were different in the microwave pyrolysis vs. conventional TGA experiments. Our results indicate that the activation energy is decreased significantly with increasing heating rates for the thermal decomposition of WPCB material. The pre-exponential factor changed with activation energy mainly as a result of a kinetic compensation effect. When WPCB material was inundated with 2.45 GHz microwaves, a rapid heating process took place, with a volumetric expulsion of volatile matter that occurs immediately after the sample exposure to microwaves. Compared with a simulated conventional TGA incorporating a similar heating rate, the activation energy in microwave-induced pyrolysis is much smaller. This can be attributed to the internal-type heating style and a catalyst effect caused by the presence of microwave heating or microwave-metal discharges. A high disposal efficiency and low activation energy indicated by the microwave-induced pyrolysis of WPCBs makes the adoption of microwave technology an attractive approach for the disposal of WPCBs and even WEEE materials.

REFERENCES

1. Schwarzer, S; de Ono, A.; Peduzzi, P.; Giuliani, G.; Kluser, S. E-Waste, the Hidden Side of IT Equipment's Manufacturing and Use; Environment Alert Bulletin; United Nations Environment Programme: Geneva, Switzerland, 2005. Available online: http://www.grid.unep.ch/products/3_Reports/ew_ewaste.en.pdf (accessed on 5 May 2005).
2. Guo, J.; Guo, J.; Xu, Z. Recycling of non-metallic fractions from waste printed circuit boards: A review. J. Hazard. Mater. 2009, 168, 567–590.
3. Byers, T.J. Printed Circuit Board Design with Microcomputer; McGraw-Hill: New York, NY, USA, 1991.
4. Das, A.; Vidyadhar, A.; Mehrotra S.P. A novel flowsheet for the recovery of metal values from waste printedcircuit boards. Resour. Conserv. Recycl. 2009, 53, 464–469.
5. Hussain, Z.; Khan, K.M.; Hussain, K. Microwave-metal interaction pyrolysis of polystyrene. J. Anal. Appl. Pyrolysis 2010, 89, 39–43.
6. Ludlow-Palafox, C.; Chase, H.A. Microwave-induced pyrolysis of plastic wastes. Ind. Eng. Chem. Res. 2001, 40, 4749–4756.
7. Miura, M.; Kaga, H.; Sakurai, A.; Kakuchi, T.; Takahashi, K. Rapid pyrolysis of wood block by microwave heating. J. Anal. Appl. Pyrolysis 2004, 71, 187–199.
8. Huang, Y.F.; Kuan, W.H.; Lo, S.L.; Lin, C.F. Hydrogen-rich fuel gas from rice straw via microwave-induced pyrolysis. Bioresour. Technol. 2010, 101, 1968–1973.
9. Du, Z.; Li, Y.; Wang, X.; Wan, Y.; Chen, Q.; Wang, C.; Lin, X.; Chen, P.; Ruan, R. Microwave-assisted pyrolysis of microalgae for biofuel production. Bioresour. Technol. 2011, 102, 4890–4896.
10. Zhao, X.Q.; Song, Z.L.; Liu, H.Z.; Li, Z.Q.; Li, L.Z.; Ma, C.Y. Microwave pyrolysis of corn stalk bale: A promising method for direct utilization of large-sized biomass and syngas production. J. Anal. Appl. Pyrolysis 2010, 89, 87–94.
11. Zhao, X.Q.; Zhang, J.; Song, Z.L.; Liu, H.Z.; Li, L.Z.; Ma, C.Y. Microwave pyrolysis of straw bale and energy balance analysis. J. Anal. Appl. Pyrolysis 2011, 92, 43–49.
12. Domínguez, A.; Menéndez, J.A.; Fernández, Y.; Pis, J.J.; Valente, J.M.N.; Carrott, P.J.M.; Ribeiro Carrott, M.M.L. Conventional and microwave induced pyrolysis of coffee hulls for the production of a hydrogen rich fuel gas. J. Anal. Appl. Pyrolysis 2007, 79, 128–135.
13. Elharfia, K.; Mokhlisse, A.; Chanâa, M.B.; Outzourhit, A. Pyrolysis of the Moroccan (Tarfaya) oil shales under microwave irradiation. Fuel 2000, 79, 733–742.

14. Menéndez, J.A.; Domínguez, A.; Inguanzo, M.; Pis, J.J. Microwave-induced drying, pyrolysis and gasification (MWDPG) of sewage sludge: Vitrification of the solid residue. J. Anal. Appl. Pyrolysis 2005, 74, 406–412.
15. Menéndez, J.A.; Domínguez, A.; Inguanzo, M.; Pis, J.J. Microwave pyrolysis of sewage sludge: Analysis of the gas fraction. J. Anal. Appl. Pyrolysis 2004, 71, 657–667.
16. Lam, S.S.; Russell, A.D.; Lee, C.L.; Chase, H.A. Microwave-heated pyrolysis of waste automotive engine oil: Influence of operation parameters on the yield, composition, and fuel properties of pyrolysis oil. Fuel 2012, 92, 327–329.
17. Lam, S.S.; Russell, A.D.; Lee, C.L.; Lam, S.K.; Chase, H.A. Production of hydrogen and light hydrocarbons as a potential gaseous fuel from microwave-heated pyrolysis of waste automotive engine oil. Int. J. Hydrog. Energy 2012, 37, 5011–5021.
18. Lam, S.S.; Russell, A.D.; Chase, H.A. Microwave pyrolysis, a novel process for recycling waste automotive engine oil. Energy 2010, 35, 2985–2991.
19. Chen, K.S.; Yeh, R.Z.; Wu, C.H. Kinetics of thermal decomposition of epoxy resin in nitrogen-oxygen atmosphere. J. Environ. Eng. 1997, 123, 1041–1046.
20. Chen, K.S.; Chen, H.C.; Wu, C.H.; Chou. Y.M. Kinetics of thermal and oxidative decomposition of printed circuit boards. J. Environ. Eng. 1999, 125, 277–283.
21. Luda, M.P.; Balabanovich, A.I.; Zanetti, M.; Camino, G. Thermal decomposition of the retardant brominated epoxy resins. J. Anal. Appl. Pyrolysis 2002, 65, 25–40.
22. Barontini, F.; Marsanich, K.; Petarca, L.; Cozzani, V. Thermal degradation and decomposition products of electronic boards containing BFRs. Ind. Eng. Chem. Res. 2005, 44, 4186–4199.
23. Chien, Y.C.; H.P.Wang; Lin, K.S.; Huang, Y.J.; Yang, Y.W. Fate of bromine in pyrolysis of printed circuit boardwastes. Chemosphere 2000, 40, 383–387.
24. Blazsó, M.; Czégény, Z.; Csoma, C. Pyrolysis and debromination of flame retarded polymers of electronic scrap studied by analytical pyrolysis. J. Anal. Appl. Pyrolysis 2002, 64, 249–261.
25. Menéndez, J.A.; Juárez-Pérez, E.J.; Ruisánchez, E.; Bermúdez, J.M.; Arenillas, A. Ball lightning plasma and plasma arc formation during the microwave heating of carbons. Carbon 2011, 49, 346–349.
26. Sun J.; Wang, W.L.; Ma, C.Y.; Dong, Y. An exploratory study of electronic waste treatment: microwave-induced pyrolysis. In Proceedings of 2010 Asia–Pacific Power and Energy Engineering Conference, Chengdu, China, 28–31 March 2010; pp. 1–4.
27. Sun J.; Wang, W.L.; Liu Z.; Ma, C.Y. Waste printed circuit boards reclamation by microwave-induced pyrolysis and featured mechanical processing. Ind. Eng. Chem. Res. 2011, 50, 11763–11769.
28. Wang, W.L.; Liu, Z.; Sun, J.; Ma, Q.L.; Ma, C.Y.; Zhang, Y.L. Experimental study on the heating effects of microwave discharge caused by metals. AIChE J. 2012, doi: 10.1002/aic.13766.
29. Sun J.; Wang, W.L.; Liu, Z.; Ma, C.Y. Study of the transference rules for bromine in waste printed circuit boards during microwave-induced pyrolysis. J. Air Waste Manag. Assoc. 2011, 61, 535–542.
30. Sun J.; Wang, W.L.; Ma, C.Y.; Dong, Y. Study on pyrolysis characteristics of electronic waste. In Proceedings of the International Conference on Chemical, Biologi-

cal and Environmental Engineering (CBEE 2009), Singapore, 9–11 October 2009; Kai, L., Ed.; World Scientific Publishing: Singapore, 2009; pp. 13–16.
31. Chiang, H.L.; Lin, K.H.; Lai, M.-H.; Chen, T.C.; Ma, S.Y. Pyrolysis characteristics of integrated circuit boards at various particle sizes and temperatures. J. Hazard. Mater. 2007, 149, 151–159.
32. Narayan, R.; Antal, M.J., Jr. Thermal lag, fusion, and the compensation effect during biomass pyrolysis. Ind. Eng. Chem. Res. 1996, 35, 1711–1721.
33. Ceamanos, J.; Mastral, J.F.; Millera, A.; Aldea, M.E. Kinetics of pyrolysis of high density polyethylene: Comparison of isothermal and dynamic experiments. J. Anal. Appl. Pyrolysis 2002, 65, 93–110.

CHAPTER 11

Studies on Pyrolysis Kinetic of Newspaper Wastes in a Packed Bed Reactor: Experiments, Modeling, and Product Characterization

APARNA SARKAR, SUDIP DE SARKAR, MICHAEL LANGANKI, AND RANJANA CHOWDHURY

11.1 INTRODUCTION

Waste management is a big issue nowadays as wastes are being generated in an ever-increasing rate by growing affluent societies. The large amount of nonhomogeneous municipal solid waste has become a tremendous problem for all Indian metropolitan cities. Municipal solid waste (MSW) consists mainly of household and commercial wastes, which are disposed of by, or on behalf of, a local authority. It is composed mainly of paper/cardboard, plastics, glass, metals, textiles, and food/garden waste. Disposal of the massive waste materials poses problems in terms of environmental impact, economic costs and technology implementation. Environment friendly processes must be thoroughly studied for the utilization of the waste materials in view of the increasing demand of energy in this modern era. The recovery of energy from the waste materials may be done through thermochemical processes like combustion, gasification, and pyrolysis. Among all these routes, pyrolysis has been receiving increasing attention in recent years as an acceptable route for waste to energy conversion. The main reason for this is that, in the pyrolysis process production of either char, oil, or gases, the pyrolysis products may be maximized by the adjustment of process condition. Pyrolysis is a thermochemical process in which hydrocarbon rich solid or liquid feed materials are thermally degraded to char, volatile liquid, and noncondensable gaseous component in absence of oxidizing media either air or oxygen. The mechanism of primary pyrolysis of solid and liquid feedstocks is as follows:

$$\text{Pyrolysing solid/liquid} \rightarrow \text{Char} + \text{Liquid (tar)} + \text{Gas (pyro-gas)} \quad \text{(I)}$$

The usual range of pyrolysis temperature is 573 K to 1273 K. All pyrolysis products have a potential use. For example, char can be burnt as fuel or disposed off (since the heavy metals are fixed inside the carbonaceous matrix), or it can even be upgraded to activated carbon [1]. Gas can be used as fuel [2], whereas oil can either serve as fuel or as a raw material for chemicals. The yield of either product, namely, char or tar or gas, may be maximized just by adjustment of operating conditions.

In recent years, a few works have been reported in the literature on pyrolysis of waste newspaper or paper mixture [3–9]. Investigation on pyrolysis of mixture of papers with various types of municipal solid wastes has been reported by a few researchers [10–14].

Data solely concerned with pyrolysis of newspaper is however lacking. According to the aforementioned literature, TGA is a common technique to study the thermal decomposition behavior as well as the chemical kinetics of thermal conversion of several biomasses. Newspaper is the principal organic solid waste of Indian metro cities. The composition of newspaper is 62% cellulose, 16% hemicellulose, and 16% lignin. Newspaper has high heating value of about 16 MJ/kg and can be converted through pyrolysis route. As the studies of kinetics of pyrolysis of a feedstock are necessary for proper understanding and application of the process, pyrolysis kinetics of newspaper has been investigated in detail under the present study. Mathematical model incorporating deactivation has also been developed.

11.2 EXPERIMENTAL

11.2.1 MATERIALS

Old newspaper samples were collected from a local residential area for pyrolysis. Table 1 summarizes the results of proximate and ultimate analyses and higher heating value of newspaper.

TABLE 1: Results of proximate and ultimate analyses and higher heating value of newspaper.

Properties	% (W/W)	Higher heating value (MJ/kg)
Moisture	10	
Volatile	75	
Ash	n.d	
Fixed carbon	15	
Carbon	59	16
Hydrogen	8.23	
Nitrogen	n.d	
Sulfur	0.31	
Oxygen (by difference)	32.44	

11.2.2 THERMOGRAVIMETRIC ANALYSIS

Thermogravimetric analysis (Pyris Diamond) of newspaper sample was conducted in the temperature range of 573 K to 1173 K, given in Figure 1. The heating rate was 10°C/min. The flow rate of N_2 gas was 150 mL/min.

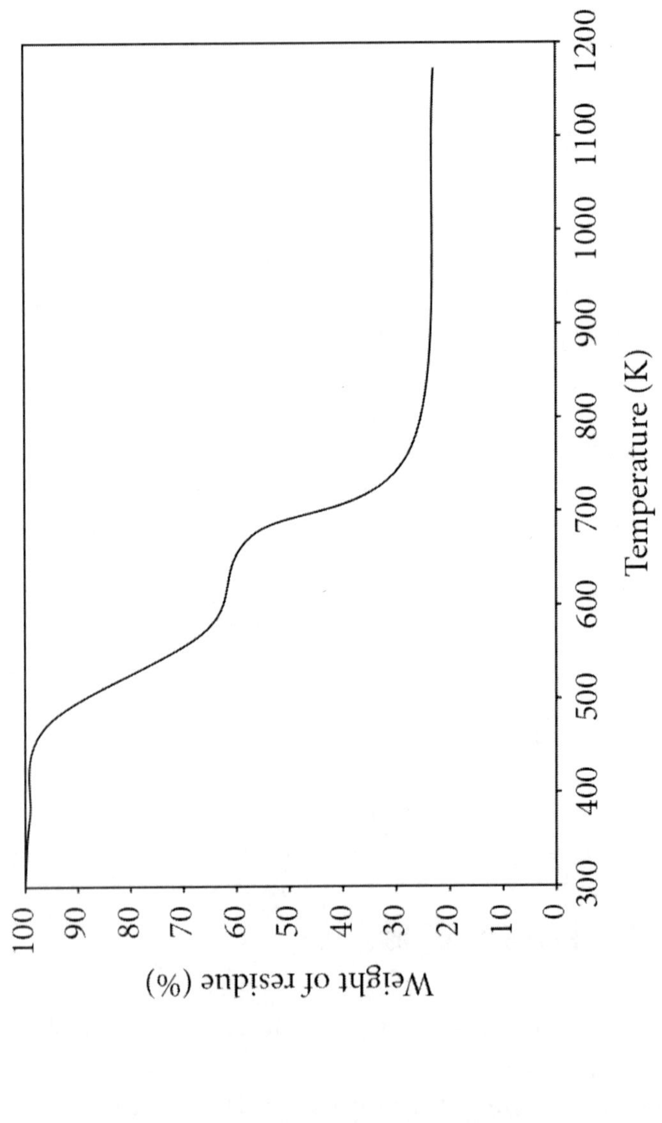

FIGURE 1: Thermogravimetric analysis graph.

From this figure, it was clear that the pyrolysis of newspaper started at above 473 K where the weight loss was approximately 20%. Pyrolysis progressed slowly from 473 K to 573 K leading to weight loss up to 35%. Above 573 K, pyrolysis became faster up to 773 K resulting in a weight loss of 74%. This was followed by slow pyrolysis up to 1173 K. The three segments of pyrolysis curve drawn by plotting weight fraction against pyrolysis temperature might signify the pyrolysis of constituent compounds of newspaper, namely, cellulose, hemicelluloses, and lignin. The fractions of cellulose, hemicelluloses, and lignin in newspaper are 40–55%, 25–40%, and 18–20%, respectively [15]. It appears that the hemicelluloses part started pyrolysing at 473 K and the main weight loss occurred for the pyrolysis of hemicellulose portion between 473 K and 573 K. Pyrolysis of cellulose part started at 573 K and proceeded up to 773 K. The last segment of the curve belonging to the temperature zone between 773 K and 1173 K signified the pyrolysis of lignin part of newspaper. Similar observation has been reported by Williams and Besler [16] during their studies on the pyrolysis of wood and its constituent component.

11.2.3 PYROLYSIS SET-UP

A 50 mm diameter and 640 mm long cylindrical stainless steel fixed bed pyrolyser was placed horizontally in a tubular furnace (Figure 2).

The pyrolyser was hung by a stainless steel chain attached with a weighing machine for continuous monitoring of the residual mass of solid in the pyrolyser. The furnace temperature varied from 573 K to 1173 K. Once the furnace temperature was raised to a preset value, pyrolyser was inserted into the furnace. Isothermal condition was maintained throughout the entire pyrolysis period. Pyrolysis was carried out for one hour at all temperatures. Experiments were designed to investigate the effects of temperature of pyrolysis on yields of pyro-oil and char and their characteristics. Nitrogen was supplied to the pyrolyser throughout the experiment to sweep the volatiles produced during pyrolysis and to maintain inert atmosphere in the reactor. The volatile product stream along with nitrogen was directed to a water cooled condenser and a series of containers placed in an ice bath. Finally, the gas stream was passed through a silica gel bed and was collected in a gas sampling bottle. The organic part of tar which got dissolved in benzene was extracted in a rotary evaporator and the quantity of pyro-oil was established. The higher heating values of the condensed pyro-oil and char samples, obtained as pyrolysis products, were determined using bomb calorimeter.

11.3 PYROLYSIS KINETICS

Pyrolysis of newspaper sample proceeds through complex reactions in series, parallel, or combination of both. Under the present study, a parallel reaction model has been attempted to describe pyrolysis kinetics of newspaper. According to this model, pyrolysis of newspaper has been considered to be a homogeneous solid phase reaction and the pyrolysis products have been considered to be char-lumped solid and volatiles made up of tar and gaseous product. The reaction pathway of pyrolysis according to the present model is as follows:

$$\text{Newspaper} \rightarrow \text{Active Complex} \xrightarrow{k_v} \text{Volatile (gas + tar)} \xrightarrow{k_c} \text{Char} \qquad \text{(II)}$$

The reaction kinetics of volatile and char has been elaborately discussed in the studies of pyrolysis of coconut shell [17], vegetable market waste [18], textile wastes [19], sesame oil cake [20], and mustard press cake [21, 22].

Figure 3 shows the experimental weight fraction profile of residue with respect to time in isothermal conditions at 573 K, 873 K, and 1173 K, respectively. From close observation of the data, it appears that the pyrolysis reactions proceed considerably in the temperature range of 573 K to 1173 K. Below this temperature range, the reactions do not occur at an appreciable rate. From the plots, it is also apparent that at each temperature, a quasiequilibrium of the reaction prevails. The rates of devolatilization reactions decline at temperatures above 673 K. Therefore, the values of frequency factors and activation energies of the reactions of reactant decomposition, volatile formation, and char formation are determined by regression analysis of the rate constant determined in the temperature range of 573 K to 673 K [17–20]. The frequency factors and activation energies of different reactions are given in Table 2.

TABLE 2: Calculated activation energies and frequency factors as per Arrhenius law.

Reaction rate constant	Frequency factor (min^{-1})	Activation energy (kJ/mol)
k	7.69	25.69
k_v	8.09	27.73
k_c	0.853	20.73

In Figure 4, the rate constants k, k_v, and k_c, calculated using activation energies and frequency factors reported in Table 2, have been plotted in the logarithmic scale against reciprocal temperature. All the rate constants calculated from the experimental results in the temperature range of 573 to 1173 K have also been superimposed on the same figure. From the figure, it is apparent that for temperatures higher than 673 K, the actual values of rate constants are far below the predictions of Arrhenius law. The deviation of the pyrolysis rate constants from the Arrhenius law may be due to the deactivation of the solid reactants with temperature. Thus, a deactivation model has been introduced. In the present study, a deactivation model has been developed on the following assumption.

1. Deactivation occurs with the increase of pyrolysis temperature beyond 673 K.
2. The apparent reaction rate constants may be written in the following form:

$$k_{ap} = ak_0 \exp\left(\frac{-E}{RT}\right)$$

$$k_{vap} = a_v kv_0 \exp\left(\frac{-E_v}{RT}\right)$$

$$k_{cap} = a_c k_{c0} \exp\left(\frac{-E_c}{RT}\right) \tag{1}$$

where a, a_v, and a_c may be defined as the activities of the solid towards corresponding reactions.

3. The values of activities a, a_v, and a_c are unity at 673 K and they decrease to the minimum of zero as the temperature increases to 1273 K.

The normalized temperature parameter has been defined as:

$$\theta = \frac{T - T(a = 1)}{T\,(a = 0) - T(a = 1)} \tag{2}$$

Here, T(a =1) = 673 K and T (a = 0) = 1273 K.

11.3.1 DEACTIVATION MODEL

In this model, the rates of deactivation are considered as function of activities themselves with exponents other than unity:

$$\frac{-da}{d\theta} = \beta a^n \quad (\text{where n}{\neq}1)$$

$$\frac{-da_v}{d\theta} = \beta_v a_v^{n_v} \quad (\text{where } n_v{\neq}1)$$

$$\frac{-da_c}{d\theta} = \beta_c a_c^{n_c} \quad (\text{where } n_c{\neq}1) \tag{3}$$

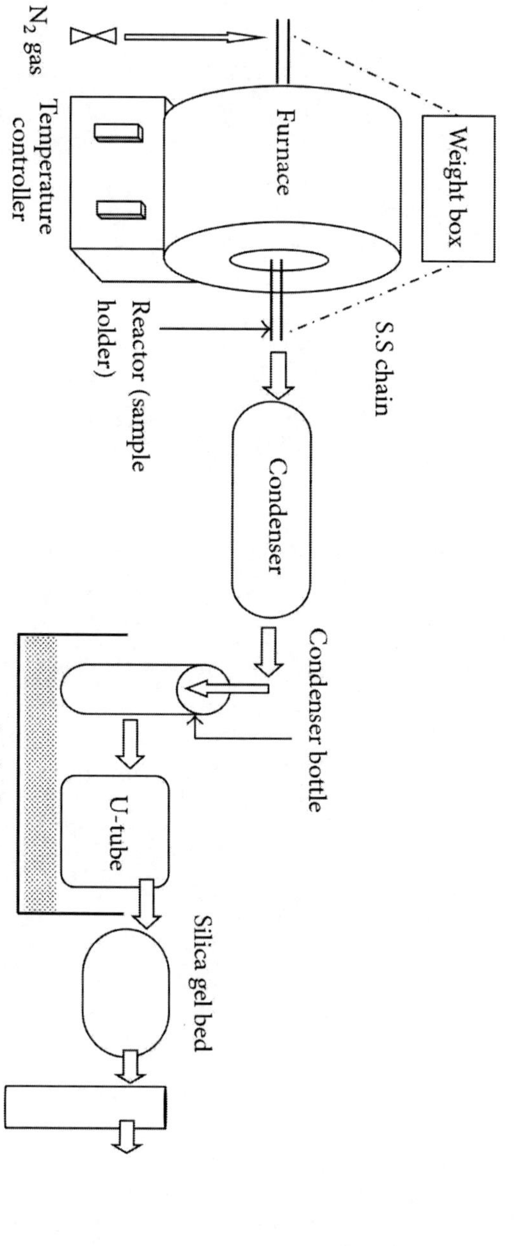

FIGURE 2: Experimental set up.

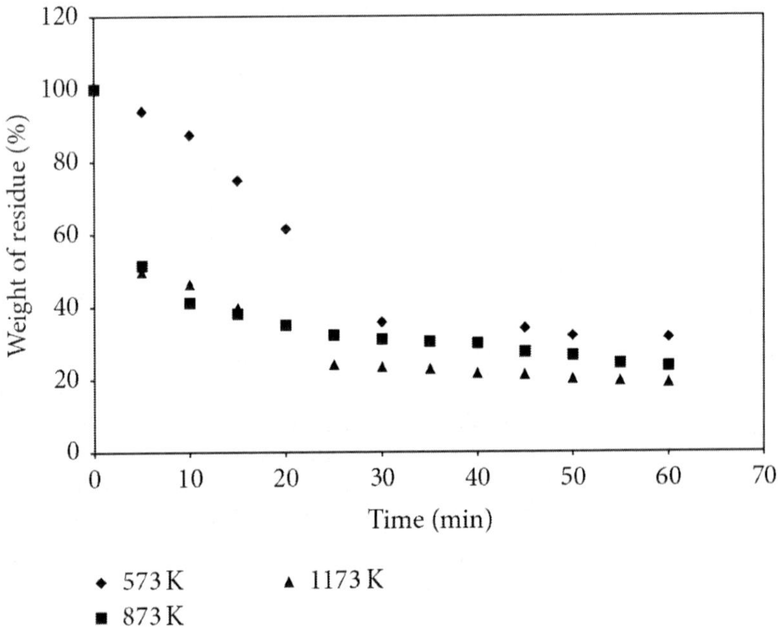

FIGURE 3: Variation of percentage of weight of residue of newspaper sample with respect to time at different pyrolysis temperature.

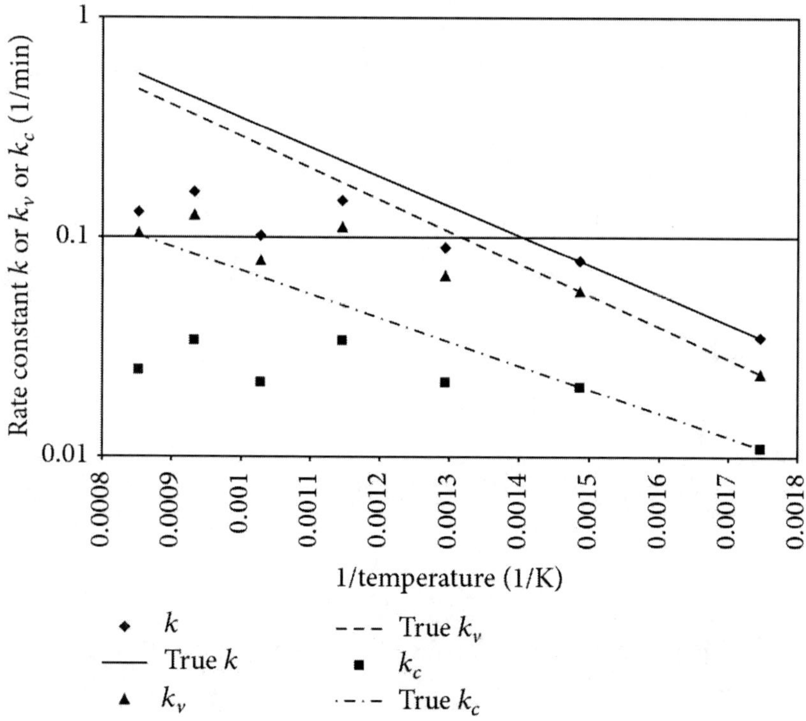

FIGURE 4: Simulated true k, true k_v, true k_c, and experimental k, k_v, and k_c values of rate constant for weight loss of newspaper, volatile formation, and char formation in logarithmic scale against reciprocal temperature.

The boundary conditions are as follows:

$$a = a_v = a_c = 0 \quad \text{at } \theta = 1$$

$$a = a_v = a_c = 1 \quad \text{at } \theta = 0 \tag{4}$$

Applying the boundary conditions represented by (4), the solutions of the differential equations may be expressed as:

$$a = \exp\left[\frac{\ln(1-\theta)}{1-n}\right]$$

$$a_v = \exp\left[\frac{\ln(1-\theta)}{1-n_v}\right]$$

$$a_c = \exp\left[\frac{\ln(1-\theta)}{1-n_c}\right] \tag{5}$$

Therefore, the rate expression may be represented as follows:

$$k_{ap} = k_0 \exp\left(\frac{-E}{RT} + \frac{\ln(1-\theta)}{1-n}\right)$$

$$k_{vap} = k_{v0} \exp\left(\frac{-E_v}{RT} + \frac{\ln(1-\theta)}{1-n_v}\right)$$

$$k_{cap} = k_{c0} \exp\left(\frac{-E_c}{RT} + \frac{\ln(1-\theta)}{1-n_c}\right) \tag{6}$$

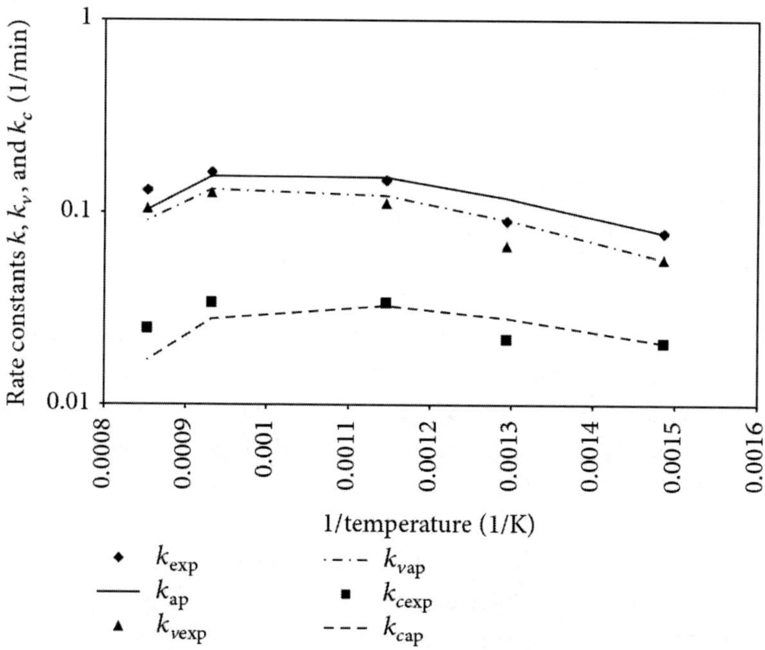

FIGURE 5: Comparison of simulated rate constants k_{app}, k_{vapp}, and k_{capp} as per deactivation model with the experimental results k_{exp}, k_{vexp}, k_{cexp} for weight loss of newspaper, volatile, and char formation, respectively.

Values of different parameter of deactivation model have been determined using nonlinear regression analysis. The values of parameters are given in Table 3.

TABLE 3: Values of constant for deactivation model.

Model parameters	Values	Correlation coefficient
n	–0.067	0.897
n_v	–0.089	0.901
n_c	–0.046	0.874

In Figure 5, logarithms of simulated values of k, k_v, and k_c, predicted by Arrhenius law using data of 573 K and 673 K as well as those predicted by deactivation model in the temperature zone beyond 673 K, have been plotted against inverse of temperature.

Comparison with experimental data suggests that for k, k_v, and k_c, the deactivation model can explain the reality, except at 973 K. The experimental findings at 973 K indicate comparatively low reaction rate. Actually, other than deactivation of single pyrolysis component due to tar clogging of active sites, conformational changes, and so forth, as observed by Bandyopadhyay et al. [17], Ray et al. [18], Sarkar and Chowdhury [19], use of different types of woods, both soft and hard, as the source of Indian newspaper may lead to abnormal pyrolysis behaviour.

11.4 RESULTS AND DISCUSSIONS

11.4.1 EFFECTS OF PYROLYSIS TEMPERATURE ON PRODUCT YIELDS

After the completion of pyrolysis of newspaper, the solid residue part was collected from the reactor. The unreacted newspaper and the left char yield were determined. While the condensable part of volatile was considered as a tar yield, the organic part of tar, soluble in benzene, was considered as pyro-oil. The gas yield was calculated by subtracting the amount of tar from volatile yield. These yields of char, tar, and gases in relation to reactor temperature are shown in Figure 6.

The char yield decreased with a rise in pyrolysis temperature from 32 wt% at 573 K to 20 wt% at 1173 K. On the other hand, the yield of tar increased to 43 wt% at 773 K and then it decreased to 12 wt% at 1173 K. Figure 6 also shows that while gas yield increased gradually from 573 K to 873 K, it increased appreciably as the pyrolysis temperature increased from 973 K to 1173 K.

Appearance of a maximum in the trend of yield of tar against temperature may be due to the commencement of further cracking of tar molecules to lower gaseous molecules at higher temperatures. Under the present experimental conditions, the results showed a clear influence of temperature on the fractional yields of char and volatiles. The decrease in the char yield with pyrolysis temperature is to be attributed to an increasing devolatilization of the solid hydrocarbons in the char. Partial gasification of the carbonaceous residue is also possible [23].

11.4.2 PRODUCTS CHARACTERIZATION

11.4.2.1 PROXIMATE ANALYSES OF CHAR

Proximate analyses of char obtained at different pyrotemperatures from 573 K to 1173 K are shown in Figure 7.

Proximate analysis of char sample has been done to measure the fixed carbon that is present in it. From this figure, it may be inferred that volatile content of char gradually decreases with temperature, while contents of fixed carbon and ash show increasing pattern with temperature. The fixed carbon and ash content of char product increased with temperature from 70.72 and 13.61 wt% at 773 K to 71.97 and 20.58 wt% at 1173 K, respectively. While volatile matter of char product decreased from 15.66 to 7.44 wt% at 773 to 1173 K.

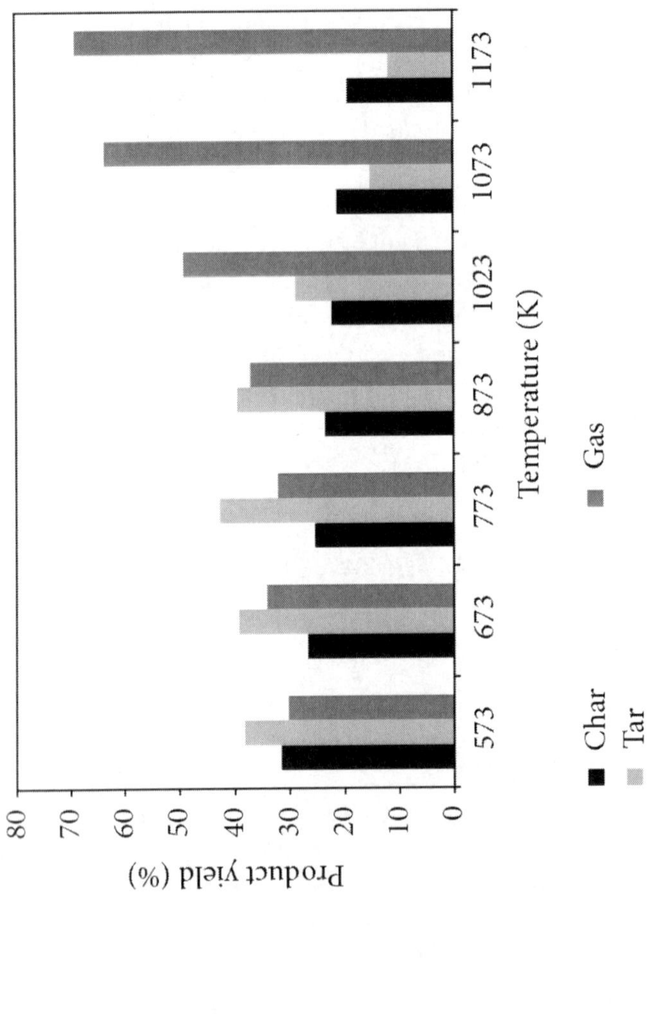

FIGURE 6: Percentage of product yields as char, tar, gas, and unreacted reactant in different pyrolysis temperature (K).

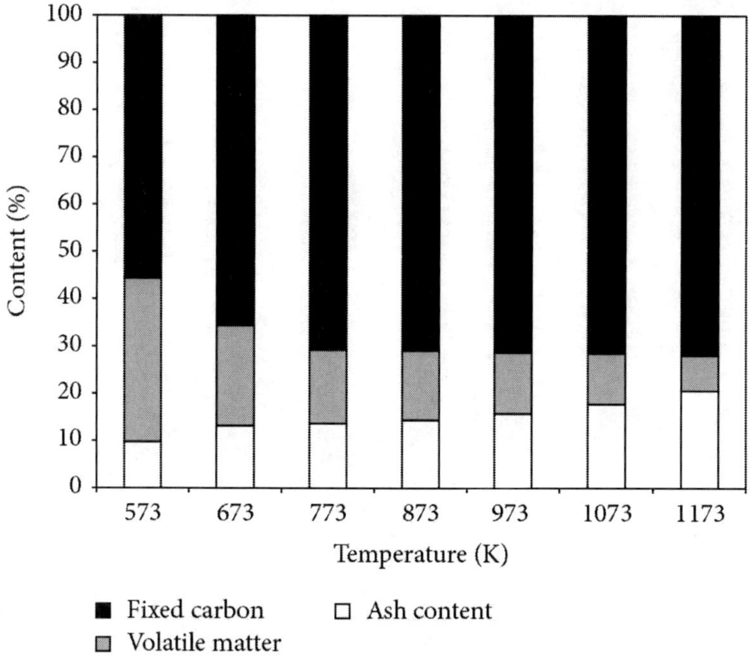

FIGURE 7: Percentage of volatile, ash, and fixed carbon present in char obtained at different pyrolysis temperatures.

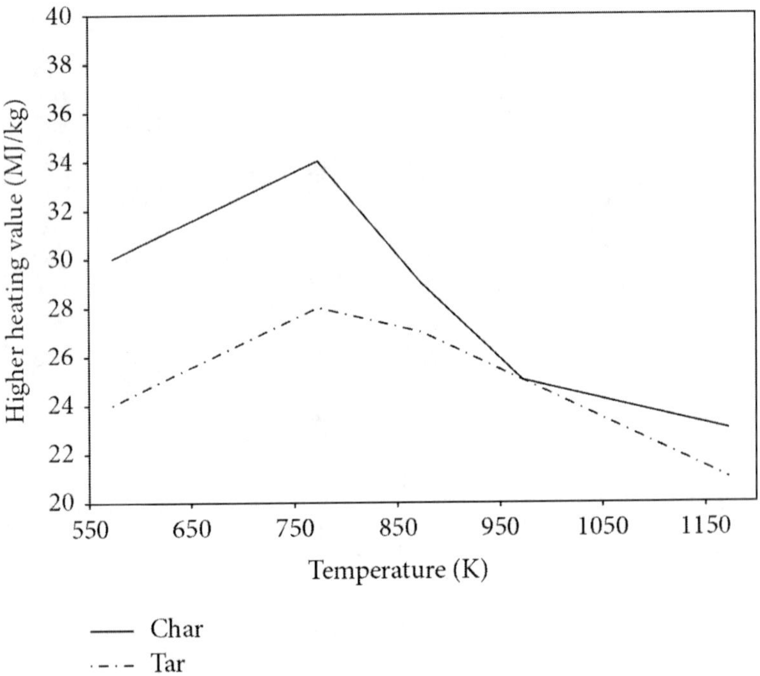

FIGURE 8: Pattern of higher heating values of pyro-oil and char at different pyrolysis temperature.

11.4.2.2 CHEMICAL CHARACTERIZATION OF PRODUCT YIELD

The empirical formulas of pyroproducts obtained at different pyrolysis temperatures are listed in Table 4. It was clear from the table that the char products became carbon rich with the rise of pyrolysis temperature. The ratio of H/C decreased with the higher temperature. On the other hand, the pyro-oil became highly oxygenated with the rise of temperature. The ratio of H/C decreased with the higher temperature. At initial temperatures, namely, 573, 673, and 773 K, the H/C ratios of pyro-oil were 14, 9, and 5, respectively. Beyond this temperature, the ratio of H/C of pyro-oil was similar to alkenes/cycloalkanes (~2).

TABLE 4: Empirical formula of char and pyro-oil.

Temperature (K)	Char yield	Tar yield
573	C_3H_2O	CH_2O
673	C_3HO	CHO
773	C_6H_3O	CHO_2
873	C_4HO	C_3HO_3
973	C_5HO	C_3HO_3
1073	C_6HO	C_3HO_3
1173	C_7HO	C_2HO_4

11.4.2.3 EFFECTS OF PYROLYSIS TEMPERATURE ON HIGHER HEATING VALUES OF PRODUCT YIELD

The higher heating values of char and tar yield of different pyrolysis temperature are determined using the bomb calorimeter given in Figure 8. The higher heating value of char increases gradually from 30 to 34 MJ/Kg as the temperature increases from 573 K to 773 K. Beyond 773 K, the higher heating value decreases from 34 to 23 MJ/Kg as the temperature is changed from 873 K to 1173 K. Pattern of temperature dependence of higher heating value of tar is also similar to that of char. This may be justified by the fact that fraction of carbon in char increases as the temperature increases up to 773 K, beyond which char further participates in heterogeneous reactions with gaseous product. In case of tar, fraction of volatile components increases with the temperature, resulting in the increase of higher heating value up to 773 K. At temperatures above 773 K, secondary cracking of tar takes place causing decrease in higher heating value of tar.

11.4.2.4 FT-IR ANALYSES OF PYRO-OIL

The FT-IR spectra of pyro-oils obtained at different temperatures from pyrolysis of newspaper waste feedstock are provided in Figure 9. Band assignments of IR spectrum of pyro-oil, which are summarized in Table 5, indicate that the pyro-oil contains a number of atomic groupings and structures.

TABLE 5: Main atomic groups and structure of pyro-oil.

Wavenumber (cm^{-1})	Infrared spectrum	Atomic groups and structures
3200–3700	O–H stretching	Hydroxyl
2800–3000	C–H stretching	Aliphatic structures
1650–1770	C=O stretching	Carbonyl
1610–1680	C=C stretching	Oliefinic structures
1450–1600	C=C stretching	Aromatic structures
1420–1480	C–H bending	Aliphatic structures
1360–1430	O–H and C–H bending	Hydroxyl, acid, phenol, olefins, and methyl
1200–1300	C–O stretching	Unsaturated ethers
1000–1200	C–H out-of-plane blending	Aromatic structures

It appeared from Figure 9 that the intensity of spectrum of pyro-oil has changed with the increase of temperature. At higher temperature, the hydrogen bonded OH stretching decreased due to the loss of phenolic or alcoholic groups of the pyro-oil [21, 24].

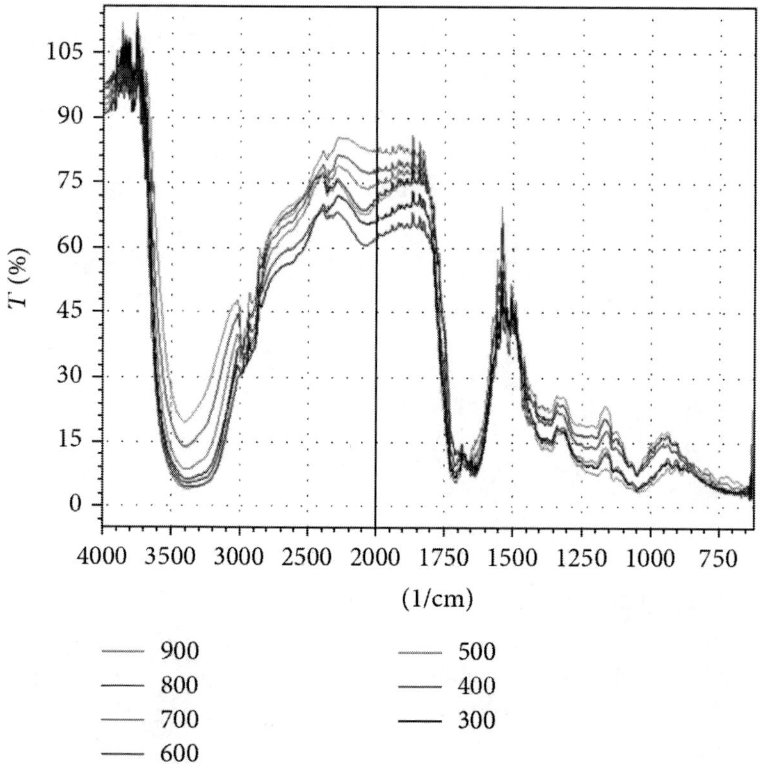

FIGURE 9: FT-IR spectrum of pyro-oil obtained at different pyrolysis temperatures.

11.5 CONCLUSION

In the present investigation, pyrolysis of newspaper has been studied in the temperature range of 573 K to 1173 K. FT-IR analyses of pyro-oil obtained at different reaction temperatures have been done. The effects of pyrolysis temperature on higher heating value and ratio of H/C of pyro-oil and char have been discussed. These properties of the pyrolysis products can be used as fundamental data for the design of a pyrolysis process for biomass wastes. The system has been mathematically modeled in a deterministic way incorporating the deactivation phenomenon. A deactivation model representing the deactivation rate as a nonlinear function of activities has been found to be successful to explain the reaction engineering behaviour of the system.

NOMENCLATURE

- a: Activity of solid
- E: Activation energy (kJ/mol)
- k: Rate constant (min^{-1})
- k_{ap}: Apparent reaction rate constant (k/min)
- R: Gas constant (kJ/mol/K)
- T: Temperature (K)
- t: Time (min)
- β: Deactivation rate constant
- θ: Dimensionless temperature parameter
- v: Volatiles
- c: Char
- 0: Initial condition.

REFERENCES

1. G. Q. Lu, J. C. F. Low, C. Y. Liu, and A. C. Lua, "Surface area development of sewage sludge during pyrolysis," Fuel, vol. 74, no. 3, pp. 344–348, 1995.

2. W. Iwasaki, "A consideration of the economic efficiency of hydrogen production from biomass," International Journal of Hydrogen Energy, vol. 28, no. 9, pp. 939–944, 2003.
3. M. N. A. Bhuiyan, K. Murahami, and M. Ota, "On thermal stability and chemical kinetics of waste newspaper by thermogravimetric and pyrolysis analysis," Journal of Environment and Engineering, vol. 3, pp. 1–12, 2008.
4. L. Li, H. Zhang, and X. Zhuang, "Pyrolysis of waste paper: characterization and composition of pyrolysis oil," Energy Sources, vol. 27, no. 9, pp. 867–873, 2005.
5. C.-H. Wu, C.-Y. Chang, C.-H. Tseng, and J.-P. Lin, "Pyrolysis product distribution of waste newspaper in MSW," Journal of Analytical and Applied Pyrolysis, vol. 67, no. 1, pp. 41–53, 2003.
6. S. Ogawa, H. Mizukami, Y. Bando, and M. Nakamura, "The pyrolysis characteristics of each component in municipal solid waste and thermal degradation of its gases," Journal of Chemical Engineering of Japan, vol. 38, no. 5, pp. 373–384, 2005.
7. C. H. Wu, C. Y. Chang, and J. P. Lin, "Pyrolysis kinetics of paper mixtures in municipal solid waste," Journal of Chemical Technology and Biotechnology, vol. 68, pp. 65–74, 1997.
8. L. Sorum, M. G. Gronli, and J. E. Hustad, "Pyrolysis characteristics and kinetics of municipal solid wastes," Fuel, vol. 80, no. 9, pp. 1217–1227, 2001.
9. M. N. A. Bhuiyan, M. Ota, K. Murakami, and H. Yoshida, "Pyrolysis kinetics of newspaper and its gasification," Energy Sources Part A: Recovery, Utilization and Environmental Effects, vol. 32, no. 2, pp. 108–118, 2010.
10. J. Zheng, Y.-Q. Jin, Y. Chi, J.-M. Wen, X.-G. Jiang, and M.-J. Ni, "Pyrolysis characteristics of organic components of municipal solid waste at high heating rates," Waste Management, vol. 29, no. 3, pp. 1089–1094, 2009. · View at PubMed · View at Scopus
11. C. H. Wu, C. Y. Chang, and J. P. Lin, "Pyrolysis kinetics of paper mixtures in municipal solid waste," Journal of Chemical Technology and Biotechnology, vol. 68, no. 1, pp. 65–74, 1997.
12. S. Luo, B. Xiao, Z. Hu, S. Liu, Y. Guan, and L. Cai, "Influence of particle size on pyrolysis and gasification performance of municipal solid waste in a fixed bed reactor," Bioresource Technology, vol. 101, no. 16, pp. 6517–6520, 2010. · View at PubMed · View at Scopus
13. M. E. Sánchez, M. J. Cuetos, O. Martínez, and A. Morán, "Pilot scale thermolysis of municipal solid waste. Combustibility of the products of the process and gas cleaning treatment of the combustion gases," Journal of Analytical and Applied Pyrolysis, vol. 78, no. 1, pp. 125–132, 2007.
14. A. N. García, R. Font, and A. Marcilla, "Kinetic study of the flash pyrolysis of municipal solid waste in a fluidized bed reactor at high temperature," Journal of Analytical and Applied Pyrolysis, vol. 31, pp. 101–121, 1995.
15. H. Jørgensen, J. B. Kristensen, and C. Felby, "Enzymatic conversion of lignocellulose into fermentable sugars: challenges and opportunities," Biofuels, Bioproducts and Biorefining, vol. 1, no. 2, pp. 119–134, 2007.
16. P. T. Williams and S. Besler, "The influence of temperature and heating rate on the slow pyrolysis of biomass," Renewable Energy, vol. 7, no. 3, pp. 233–250, 1996.

17. S. Bandyopadhyay, R. Chowdhury, and G. K. Biswas, "Thermal deactivation studies of coconut shell pyrolysis," The Canadian Journal of Chemical Engineering, vol. 77, no. 5, pp. 1028–1036, 1999.
18. R. Ray, P. Bhattacharya, and R. Chowdhury, "Simulation and modeling of vegetable market wastes pyrolysis under progressive deactivation condition," Canadian Journal of Chemical Engineering, vol. 82, no. 3, pp. 566–579, 2004.
19. A. Sarkar and R. Chowdhury, "Reaction kinetics and product distribution of slow pyrolysis of Indian textile wastes," International Journal of Chemical Reactor Engineering, vol. 10, no. 1, pp. 1–22, 2012.
20. A. Sarkar, B. Mondal, and R. Chowdhury, "Mathematical modeling of a semibatch pyrolyser for sesame oil cake," Industrial and Engineering Chemistry Research, vol. 53, no. 51, pp. 19671–19680, 2014.
21. A. Sarkar, S. Dutta, and R. Chowdhury, "Mustard press cake pyrolysis and product yield characterization," International Journal of Scientific & Engineering Research, vol. 4, no. 8, 2013.
22. A. Sarkar and R. Chowdhury, "Studies on catalytic pyrolysis of mustard press cake with NaCl," International Journal of Engineering Sciences & Research Technology, vol. 3, pp. 90–96, 2014.
23. J. A. Menéndez, A. Domínguez, M. Inguanzo, and J. J. Pis, "Microwave pyrolysis of sewage sludge: analysis of the gas fraction," Journal of Analytical and Applied Pyrolysis, vol. 71, no. 2, pp. 657–667, 2004.
24. R. K. Sharma, J. B. Wooten, V. L. Baliga, X. Lin, W. G. Chan, and M. R. Hajaligol, "Characterization of chars from pyrolysis of lignin," Fuel, vol. 83, no. 11-12, pp. 1469–1482, 2004.

CHAPTER 12

A Review on Waste to Energy Processes Using Microwave Pyrolysis

SU SHIUNG LAM AND HOWARD A. CHASE

12.1 INTRODUCTION

Waste materials such as waste oil, plastic, and biomass waste are being generated every year around the world. Some of these wastes are effectively collected and recovered for use as an energy source or chemical feedstock, while some are simply discarded or burned in ways that can pollute the environment. The improper disposal of these waste materials (e.g., used engine oil, PVC, and municipal solid wastes) may constitute an environmental hazard due to the presence of undesirable species such as metals, soot and polycyclic aromatic hydrocarbons (PAHs). For example, studies have reported that the high concentrations of PAHs in sewage effluents and urban runoffs are due to their contamination by waste oil [1]. Due to the difficulties associated with the contaminants in these waste materials, a large portion of the wastes are simply disposed of by landfilling. This method presents no recovery of the potential value of the waste but leads to potential environmental pollution due to considerably long decomposition times and the risk of contaminants leaking out to the surrounding environment. As such, the treatment and disposal of these wastes represents a significant challenge in the real world.

Another form of waste disposal is achieved by thermal means through incineration, which is typically used for energy recovery purposes. In this energy recovery process, waste materials are burned in furnaces to produce energy in the form of electricity and/or heat. This method can be applied to all types of hydrocarbon waste; however, it has several drawbacks. The combustion of waste results in the release of greenhouse gas such as CO_2 that contributes to climate change. In addition, incineration can lead to toxic emissions that pose a direct hazard to the environmental and human health [2,3], largely due to the undesirable contaminants present in both the waste (e.g., soot or particulates containing PAHs, metals from engine oil) and the flue gas emitted from the incineration process (e.g., fly ash, oxidised compounds such as polychlorinated dibenzodioxins). On the whole, incineration recovers only the calorific value of the waste and does not allow for the recovery of any of the chemical value of the waste, and thus this method is becoming increasingly impracticable due to concerns of greenhouse gas release and environmental pollution associated with toxic emissions. Furthermore, the cleaning of the flue gases produced is complex and expensive due to strict regulations on atmospheric emissions. Therefore, research has moved towards developing a better solution, from an environmental and economic standpoint, by thermally reprocessing the waste materials into more useful energy forms using thermal conversion processes.

Thermal conversion involves the use of a wide range of thermal decomposition processes such as pyrolysis and gasification to decompose waste materials into smaller molecules that can be used as energy source or inputs for the synthesis of new materials, e.g., hydrocarbon wastes are decomposed to produce syngas (H_2 + CO); the syngas can be used as a fuel directly, or converted into liquid fuel through the Fischer–Tropsch process [4]. In these processes, waste materials are heated and cracked in the absence of oxygen into smaller molecules. While thermal conversion shows advantages in dealing with a wide variety of wastes of a hydrocarbon nature (e.g., used tyres [5] and plastic waste [6]), and recovering both the energy and chemical value of the wastes [7], it is an energy intensive process that usually requires a large scale operation coupled with a capital intensive plant [8]. Although thermal conversion techniques have recently shown potential as an environmentally friendly waste disposal method [9,10], such practice is yet to become popular.

Pyrolysis is a thermal process that heats and decomposes a substance in an environment from which oxygen is excluded. It can be used as a thermal conversion technique for hydrocarbon wastes, where the waste materials are cracked to produce hydrocarbon oils, gases, and char. The process can be optimised to maximise production of any of these constituents by altering parameters such as process temperature and reactant residence time [11], e.g., a high temperature and high residence time promotes the production of gases; a high temperature and low residence time (termed "flash pyrolysis") results in increased yield of condensable products, and a low temperature and heating rate leads to increased char production [12] or to no chemical reactions taking place at all. The wide variety of pyrolysis products indicates that the products may need to be separated and purified before they can be used further; this can usually be achieved through the use of existing distillery and refinery facilities, however, in some cases more advanced techniques are required for the separation, e.g., the use of material separation agents for the separation of azeotropes in pyrolysis oil product (termed "azeotrope or extractive distillation").

The review focuses on the development of an emerging pyrolysis technology termed microwave pyrolysis employed for efficient treatment and energy recovery from various waste materials, either as a disposal route to convert waste materials to petrochemical products suitable for use as a fuel, or as an integrated treatment process for the production of raw chemical feedstock for future reuse. This technology is expected to offers a number of advantages over other conventional pyrolysis processes and makes a practical contribution to the challenge of redesigning human behaviour for sustainability.

12.2 CURRENT PYROLYSIS TECHNIQUES FOR WASTE TO ENERGY APPLICATION

Pyrolysis techniques have been developed as an alternative to treat and convert waste materials to products suitable for use as a potential energy source [10,13,14], though the use of this technology is not currently widespread. The main advantage of pyrolysis is that it has the potential to recover both the energy and chemical value of the waste by generating potentially valuable products from the pyrolysis process. The oil and gaseous products demonstrated a high calorific value, and the char produced can be used as a substitute for carbon black. In particular, the gaseous product is of considerable interest due to its potential as a source of hydrogen fuel. Other advantages compared with steam reformation processes include negligible production of toxic oxidised species (e.g., dioxins, NOx) [15,16], less energy consumption and the production of a disposable solid waste (char) [17]. Due to its ability to produce potentially valuable products, vigorous efforts have been made to perfect the pyrolysis process and techniques for energy recovery from waste materials, in addition to offering an alternative solution to disposal of the waste by incineration.

For the past two decades, research on pyrolysis processes has been conducted using several types of equipment heated by conventional heating source (e.g., an electric or gas heater), namely: fluidised bed reactors, rotating cone reactors, melting vessels, blast furnaces, tubular or fixed bed reactors [14,18]. For example, the institutions in Spain and Turkey use tubular and fluidised bed reactors heated by either electric furnaces, ovens, or heaters [5,19–21], whereas the Korean institutions employ stirred batch reactors heated by either jacketed electric heaters [22], autoclaves, or molten salt baths [23]. These types of equipment were used in a manner where the thermal energy is externally applied to the reactor and heats all the substances in the reactor including the evolved pyrolysis-volatiles, the surrounding gases, and the reactor chamber itself. In this case, energy is not fully targeted to the material being heated and this results in significant energy losses in terms of the energy efficiency of the whole process. Nevertheless, several of these processes have been developed into a pilot plant scale despite their limited energy efficiency [24].

Song et al. [9] have recently examined the use of electric arc heating to pyrolyse waste oil. The electric arc pyrolysis employs a different heating mode compared to conventional electric-heated pyrolysis. An electric arc cell or generator was used to generate a momentary electric discharge for pyrolysing the waste oil in order to produce high-value fuel gases (e.g., hydrogen and acetylene) and "usable" carbonaceous residue. It is considered a flash pyrolysis process whereby the waste material is rapidly heated to 1300–1500 °C for 0.01–1 s, producing incondensable gases and carbonaceous residues as the pyrolysis products. Although this technique shows potential as a waste-to-energy method, such technique should be investigated further due to concerns over the presence of undesirable species (e.g., PAHs, metals, and mixed oil-additive polluted residues) in the pyrolysis products [25].

It was established that current pyrolysis techniques (mostly conventional electric-heated pyrolysis processes), when compared with incineration and steam reformation processes, offer a number of advantages and shows excellent potential for waste-to-energy applications. However, such practices possess limitations and at present there are still problems associated with these pyrolysis techniques. The low thermal conductivity of some waste materials (e.g., engine oil with a thermal conductivity ranging from 0.15 to 0.30 W/m.K) often necessitates a long processing time due to the low rate of heat transfer within the material in order for pyrolysis to occur. Furthermore, in conventional electric-heated pyrolysis, the waste material is heated by an external heating source which also heats all the substances in the heating chamber including the evolved pyrolysis-volatiles and the chamber itself. This results in significant energy losses and can also promote undesired secondary reactions of the evolved pyrolysis-volatiles that lead to formation of toxic compounds (e.g., PAHs) [26] and increased production of char [17], which can cause problems such as coking on the reactor wall and fouling of the system with particulates. Moreover, the uneven distribution of heat produced in some conventional pyrolysis processes has led to poor control over the heating process; as a result, the final fractions obtained from the pyrolysis are often varied and critically depend on the actual process conditions applied to the waste material [27]. It was also found that the existing literature is limited to pyrolysis performed in batch or semi-batch operation in which the waste material was added initially to a batch system before being subjected to pyrolysis. Limited information is therefore available concerning the characteristics of the pyrolysis of the waste material (e.g., the influence of key process parameters on the product distribution) when the pyrolysis is performed in a continuous operation. Owing to the limitations, inconsistent performance, and uncertainties shown by conventional pyrolysis, it is important to find an alternative pyrolysis technique to rectify these deficiencies in order to ensure better performance and control of the pyrolysis process as well as the production of more desirable pyrolysis products.

12.3 MICROWAVE PYROLYSIS

Microwave pyrolysis is a relatively new process and was initially developed by Tech-En Ltd. in Hainault, UK [28,29]; this thermal treatment in a microwave-heated bed of particulate-carbon has been shown to be an effective method of recovering and recycling chemicals present in troublesome wastes such as plastic waste, sewage sludge, and coffee hulls [7,12,30,31]. In this process, waste material is mixed with a highly microwave-absorbent material such as particulate-carbon, which absorbs microwave energy to generate sufficient thermal energy to achieve the temperatures required for extensive pyrolysis to occur. As a result of microwave heating, the waste material is thermally cracked in the absence of oxygen into smaller molecules. The resulting volatile products are either recondensed into an oil product (termed "pyrolysis-oil") or collected as incondensable gaseous products (termed "pyrolysis-gases") of different compositions depending on the reaction conditions.

12.3.1 PRINCIPLES BEHIND MICROWAVE PYROLYSIS

12.3.1.1 INTRODUCTION TO MICROWAVE HEATING

Microwave heating is performed at frequencies of 915 MHz ($\lambda = \sim 33$ cm) and 2.45 GHz ($\lambda = \sim 12$ cm) as specified by international agreement [32]. The technique was introduced after the Second World War following the development of the magnetron valve (a very high power source of microwaves with exceptional efficiency) as an efficient processing tool for heating. The heating occurs when the magnetron generates electromagnetic radiation that causes dipolar molecules to attempt to rotate in phase with the alternating electric field created by the electromagnetic radiation. Resistance to this rotation on the molecular level results in friction between the molecules and causes heat to be generated [33]. This heating method has many advantages over conventional thermal heating methods.

Conventional thermal heating usually employs an external heating source to transfer heat to material through a surface; as a result, heating is governed by the temperature of that surface as well as limited by the physical properties of the materials being heated, such as density, heat capacity, and thermal diffusivity of the material. In contrast, microwave heating constitutes a unique way of heating where the heating effect arises from the interaction of electromagnetic wave with the dipoles within the material being heated. In general, three mechanisms are responsible for the microwave heating resulting from the interactions mentioned above (see Section 3.1.2). By such heating mechanisms, heat is generated within the material rather than from an external source, thereby giving a more efficient heating process compared to conventional surface heating with respect to even distribution of heat and easier control over the heating. In addition, high temperatures and heating rates can be obtained through microwave heating [7], and it shows remarkably high conversion efficiency of electrical energy into heat (80%–85%) [15].

12.3.1.2 MICROWAVE HEATING MECHANISMS

Microwave heating is classified as an electric volumetric heating method. Other heating methods in this category include conduction and induction heating operating at frequency ranges of 0–6 Hz and 50 Hz–30 kHz, respectively; these heating methods were performed by passing a current through the workload to induce electric power (I2 R) heating. Ohmic heating, which operates at the frequency range in between conduction and induction heating, is also classified into the same category and this heating method operates with the same principle as the former methods. Radio frequency heating, operating at a frequency range from 1 to 100 MHz (often at 27.12 MHz), is included in this category and this heating method is usually used for workloads that give a high resistivity when placed between electrodes. A typical electromagnetic spectrum with examples of applications performed by different electric volumetric heating methods at various frequency ranges is presented in Figure 1.

The principles and theories behind microwave heating are well understood and available in several textbooks [32,34]. In general, there are three mechanisms by which materials are heated in a microwave field. These mechanisms arise from the displacement that charged particles in the material undergo when they are subjected to the microwave radiation and could be summarised as follows:

1. Dipole reorientation (polarization)—substances containing polar compounds are mainly heated via this mechanism when they are subjected to microwave radiation. When subjected to a microwave field, the electrons around the nuclei (electronic polarization) or the atomic nuclei themselves (atomic polarization) are displaced from their equilibrium position, forming induced dipoles. In some materials (e.g., water) permanent dipoles exist due to the asymmetric charge distribution in the molecule. The induced or permanent dipoles tend to reorient under the influence of the changing or alternating electric field. The chemicals bonds of induced or permanently polarized molecules are realigned in the fluctuating field. This realignment occurs trillions of times each second [35] and results in friction between the rotating molecules, and thus causes heat to be generated within the whole volume of material.
2. Interfacial or Maxwell-Wagner polarization—this polarization arises from a charge build-up in the contact areas or interfaces between different components in heterogeneous systems. The polarization is created due to the difference in conductivities and dielectric constants of the substances at the interfaces. The accumulation of space charge leads to field distortions and dielectric loss that contribute to the heating effects.
3. Conduction mechanism—when an electrically-conductive material is subjected to electromagnetic radiation, electric currents are

produced where the charged particles or carriers (electrons, ions, etc.) in the material move through the material under the influence of the externally applied electromagnetic field forming conducting paths. As these electric currents flow within the structure of the materials, which in most cases have a relatively high electrical resistivity, the material is heated because the power generated by the forced flow of electrons is dissipated as heat.

The extent to which a material heats up when subjected to electromagnetic radiation (e.g., microwave radiation) is mainly determined by its dielectric properties, which depends on two key parameters: the dielectric constant (ε') and the dielectric loss factor (ε"). The dielectric constant expresses the ability of a material to be polarized by an electric field by determining how much of the electromagnetic energy is reflected and how much is absorbed, whereas the dielectric loss factor quantifies the efficiency with which the electromagnetic energy is converted to heat. The ratio of dielectric loss factor to dielectric constant defines the dielectric loss tangent or dissipation factor of a material:

$$\tan \delta = \varepsilon''/\varepsilon' \quad (1)$$

The dielectric loss tangent of a material determines its ability to absorb and convert electromagnetic energy into thermal energy at a given temperature and frequency. In microwave heating process, this parameter is an important factor that determines the heating rate and final temperature that can be reached by a material heated by microwave radiation [36]. Thus, a material with a moderate value of ε' and a high value of ε" (and so a high value of tan δ) is considered as a good microwave receptor with high capability in converting electromagnetic energy into thermal energy. Typical materials that exhibit good dielectric properties with a high value of tan δ are carbon materials and inorganic oxides [37], whereas materials such as plastics are considered as 'transparent' to microwaves because they do not possess a sufficiently high dielectric loss factor (ε") to allow for dielectric heating [35].

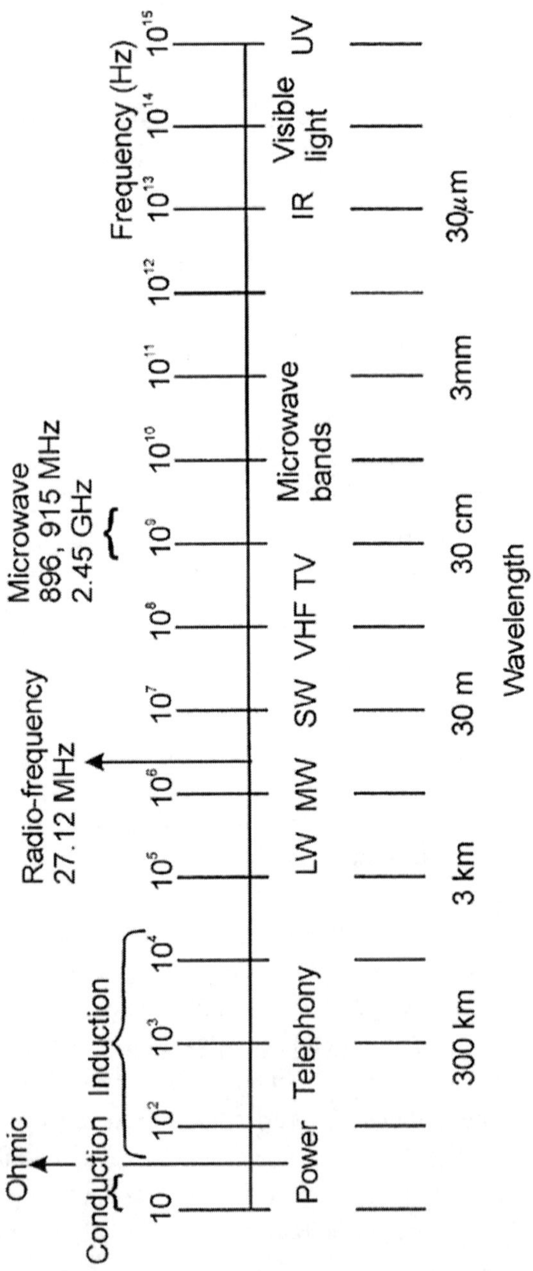

FIGURE 1: Typical electromagnetic spectrum with examples of applications performed at different frequency ranges [32].

12.3.1.3 MICROWAVE HEATING OF CARBON MATERIALS AND ITS APPLICATION IN PYROLYSIS PROCESSES

Microwave heating combined with the use of carbon material has been applied in the processing and treatment of certain materials. Carbon materials are good microwave-absorbents that show high capacity to absorb and convert microwave energy into heat; the dielectric loss tangent (tan δ) of carbon materials such as charcoal, carbon black, and activated carbon, which range between 0.1 and 0.8, is either comparable to or higher than the tan δ of distilled water (~0.1), which is commonly known as a very good microwave-absorbent [35]. Thus, carbon materials can be used as an effective microwave-absorbent to heat substances that are transparent to microwaves radiation. The use of carbon materials combined with microwave heating has been applied to soil remediation processes and catalytic heterogeneous reactions [35].

Microwave heating combined with the use of carbon materials has recently been applied in pyrolysis processes to treat or process a variety of materials, e.g., biomass [12,38], coal [39,40], oil shale [41], glycerol [42] and various organic wastes [28]. In general, these materials are poor microwave-absorbent that are either transparent to microwaves or with poor dielectric properties, therefore they requires heating by contact with materials with high microwave absorbency (e.g., carbon materials [12,38,41–43] or metal oxides [39]) to achieve higher temperatures in order for extensive pyrolysis to occur. Carbon materials are usually used rather than metal oxides due to their low cost and ease of acquisition [35]. This type of pyrolysis (termed "microwave pyrolysis"), which involves the use of microwave radiation as an indirect heat source combined with the use of carbon materials as the microwave receptor to directly heat and pyrolyse the materials, is known to offer additional advantages over conventional pyrolysis techniques (see Section 3.2).

The heating of carbon occurs when sufficient microwave energy is absorbed by the carbon particles and converted to thermal energy by dipole reorientation and ionic conduction (the 1st and 3rd mechanism described in Section 3.1.2), and thus causing heat to be generated within the particles

[44]. A detailed study of the ionic conduction mechanism has been presented and discussed by Liu et al. [45]. The group led by Menendez has recently demonstrated that the microwave heating of carbon materials also occurs via the combination of Maxwell–Wagner polarization and conduction mechanism (the 2nd and 3rd mechanisms described in Section 3.1.2) [35]. The authors claimed that a current travelling in phase with the electromagnetic field is induced within the carbon material when it is subjected to microwave radiation. As a result, the π-electrons (i.e., charged particles that are free to move in a limited region within the material) in the carbon material are displaced from their equilibrium positions and this leads to dielectric polarization. The positive charges are displaced toward the field and negative charges are shifted in the opposite direction under the influence of dielectric polarization. This creates an internal electric field which reduces the overall field within the carbon material itself. As the π-electrons are repeatedly shift from one position to another under the influence of the changing or alternating electric field, this leads to a charge build-up within the carbon material. The power generated by the forced flow of electrons and the accumulation of charge within the carbon material due to Maxwell-Wagner polarization (which leads to field distortions and dielectric loss) result in energy being dissipated as heat and thus contributing to the heating effects. Figure 2 shows a schematic representation of the microwave heating of carbon material proposed by the Menendez group.

In addition, these authors demonstrated that an additional phenomenon may occur in microwave heating of carbon materials. They claimed that in some cases the kinetic energy of some π-electrons in the carbon material may increase to a level that allows the π-electrons to jump out of the carbon, resulting in the ionization of the surrounding atmosphere [35]. This phenomenon is perceived as sparks or an electric arc at a macroscopic level, but it is considered as hot spots or plasmas at a microscopic level. The generation of sparks or hot spots may have effects on the reaction pathways that occur in processes that incorporate the microwave heating of carbon.

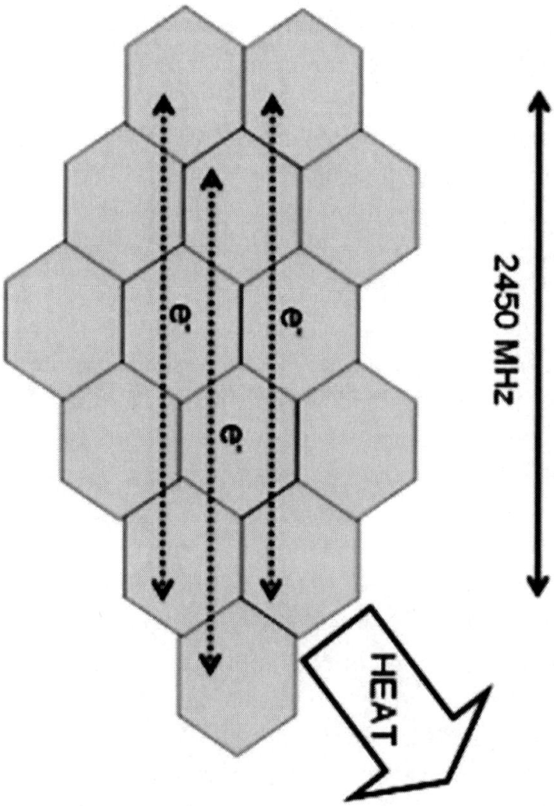

FIGURE 2: Schematic representation of microwave heating of carbon material [46].

12.3.2 MICROWAVE-HEATED PYROLYSIS COMPARED TO CONVENTIONALLY-HEATED PYROLYSIS

Microwave heating is currently employed for energy transfer in various industrial, technological and scientific processes and applications. Industrial microwave heating systems have been developed for a variety of applications in food industry, e.g., continuous baking, vacuum drying, tempering and thawing, pasteurisation, and sterilisation [32]. In addition, this technology has long been employed in the rubber and polymer industries for applications such as rubber vulcanization, polymerization or curing of resins and polymers by elimination of polar solvents [35], but there is a growing interest for its potential use as a heat source in pyrolysis processes (termed “microwave pyrolysis”), particularly in the treatment of various waste-streams.

It has been reported that microwave-heated processes showed advantages over their conventionally heated counterparts (e.g., conventionally heated oil baths, conventional electric-resistance heated systems such as electric-heated mantles and furnaces) in promoting clean, fast and high yielding chemical reactions that occur during the thermal processes [47,48]; the chemical reactions refer mainly to organic reactions, though the interpretation to some extent is also relevant to other thermal processes involving organometallic and inorganic compounds, and materials such as ceramics and polymers; in addition, the reaction is described as “clean” due to its low hold-up of potentially dangerous intermediates and its significantly lower production of contaminants. The authors claimed that these beneficial effects could be attributed to the lower thermal inertia and faster response (e.g., increasing reaction rates by reducing activation energies) exhibited by microwave heating methods. Despite the various studies to establish the difference between microwave-heated and conventionally-heated processes, Strauss and Rooney [49] have recently demonstrated that the claims are valid only when comparably rapid heating cannot be obtained by conventional heating.

Although reports that microwave-heated reactions typically proceed faster, more cleanly, and in higher yields than their conventionally heated counterparts have widely appeared in literature [50,51], the explanation behind such findings has been interpreted as mostly speculations due to the lack of compelling scientific evidence [52,53]. Some workers had tried to attribute non-thermal effects as the source of the faster, cleaner and higher yielding reactions occurred in microwave-heating processes [31,54], however a number of studies have demonstrated that the vast majority of the reactions proceed thermally under microwave-heated conditions [55,56]. These studies indicated that the reactions occurred under microwave-heated conditions should produce identical outcomes to those occurred under conventionally-heated conditions if their thermal profiles are identical [57]. Nevertheless, it can be established from the many literature reports on microwave-heated processes that microwave heating is generally accepted to have an influence in promoting and accelerating certain chemical reactions due to the advantageous features such as fast heating rates, high power density, even heating, and good heating control.

Despite the vigorous debate about the difference between microwave-heated and conventionally-heated processes, it should be noted that the discussion in the previous paragraph focuses mainly on materials heated directly by microwave radiation in order to produce physicochemical changes. In contrast, this review describes the work performed to study a microwave-heated process that employs a different approach by using microwave radiation as an indirect heat source to heat and pyrolyse waste materials. In this microwave-heated process (termed "microwave pyrolysis"), waste material is pyrolysed by contact with a bed of particulate-carbon heated by microwave radiation. The highly microwave-absorbent particulate-carbon absorbs enough microwave energy and heats up initially to achieve the desired temperature for pyrolytic thermal cracking, and subsequently crack the waste material to produce a variety of different products. Thus, this microwave pyrolysis process, which involves the preheating of carbon followed by the transfer of heat to the waste material in order for cracking to occur, may present a different concept (e.g., heating mechanism) and produce a different outcome as to the microwave-heated processes described in the previous paragraph.

It was established that microwave pyrolysis offers a number of advantages over conventional pyrolysis techniques. The use of microwave radiation as a heat source is known to offer additional advantages over traditional thermal heat sources [58,59], and the combination of carbon-based material and the novel use of microwave heating in pyrolysis processes is of increasing interest as reflected by considerable recent research reported in the literature [30,42]. Microwave radiation provides a rapid and energy-efficient heating process as compared to conventional technologies (e.g., 50% more efficient than heating by natural gas, steam) [60]. The diffuse nature of the electromagnetic field allows microwave heating to evenly heat many substances in bulk [61], without relying on slower and less efficient conductive or convective techniques. In addition, microwave heating offers a reliable, low cost, powerful heat source, with modern equipment operating at over 90% conversion efficiencies of electricity into thermal energy [15]. Thus, the use of microwave radiation as a heat source offers an improved uniformity of heat distribution, excellent heat transfer, and provides better control over the heating process. Other advantages compared with conventional heating techniques (e.g., electrical furnace) include fast internal heating [61], higher power densities and the ability to reach high temperatures at faster heating rates [7], higher heating efficiency [58], facilitating increased production speeds and decreased production costs [62]. The process can be physically gentle, allowing for a wide variety of applications in diverse fields [58,60]; increased process yield, environmental compatibility, and savings in process time are among the advantages reported on microwave processing of materials [63]. Furthermore, the process can be developed at a variety of scales that allows for on-site treatment of waste material, and thus reducing the additional costs needed for the transport of waste material to a specific treatment centre or waste processing plant.

Microwave radiation can be used to directly heat substances if they exhibit good dielectric properties; substances that are 'transparent' to microwaves (e.g., plastics), or with poor dielectric properties (e.g., engine oil) necessitate the introduction of an appropriate intermediate microwave receptor. It was revealed that the use of carbon-based materials (e.g., particulate-carbon, char produced from pyrolysis) as the microwave receptor offers a number of advantages over conventional pyrolysis techniques. Carbon-based materials are good microwave-absorbents [37] that can be used to heat neighbouring substances (particularly those that are transparent to microwaves) to high temperatures at a fast heating rate by microwave radiation [38,61,64]. In addition, the use of carbon-based materials as a reaction bed provides a highly reducing chemical environment, which removes oxygen functionalities from the treated substances and decreases the formation of undesirable oxidised species during the pyrolysis [7,44]. Short heat transfer distances, the enveloping nature of the well-mixed carbon bed, and small particle size of the carbon materials employed (with corresponding large surface area) make this an efficient method of heat transfer. Furthermore, energy is targeted only to microwave receptive materials and not to gases within the heating chamber or the chamber itself. It can promote certain chemical reactions in a way that is not possible in conventional processing by selectively heating the reactants, leading to a more uniform temperature profile and improved yield of desirable products [30,37,59,60].

Despite the many advantages shown by microwave pyrolysis, it should be noted that microwave radiation presents an additional hazard over traditional heating methods, although this is easily contained within an appropriate Faraday cage. Also, using microwaves places limits on which materials can be used in the construction of a reactor and its design. In particular, metal should be avoided other than to contain the microwaves, as the presence of metal can generate arcing that may cause damage to the microwave equipment.

A Case Study Comparison between Microwave-Heated and Conventional Electric-Heated Pyrolysis of Waste Oil

Table 1 compares the results obtained using microwave pyrolysis process to those of waste oil pyrolysis processes heated by conventional electric heating either using waste oil on its own or in the additional presence of coal, scrap tyres, or zeolite and alumina catalysts. The use of the microwave-heated bed of particulate-carbon, compared to the other methods of operation, seemed to have a beneficial effect in cracking the waste oil to produce higher amounts of condensable products (termed "pyrolysis-oil"). In addition, the examination of the hydrocarbon composition of the pyrolysis-oil revealed that waste oil was thermally cracked to mainly C5-C18 hydrocarbons (Table 1) compared to the heavier C19-C35 hydrocarbons produced in conventional electric-heated pyrolysis [5,10,22]. These results suggest that cracking reactions are enhanced through the use of a microwave-heated bed of particulate-carbon in the experimental set-up, transforming the waste oil into (greater amounts of) pyrolysis-oil comprising higher amounts of lighter hydrocarbon components.

TABLE 1: Comparison of product yield (wt%) in waste oil pyrolysis processes heated with different media, and driven by microwave heating and conventional electric heating (wt%—weight percentage).

Type of waste oil pyrolysis	Char	Pyrolysis-gases	Pyrolysis-oil
Microwave heating with particulate-carbon [65]	7	8	85
Electric heating (only waste oil) [10,19,27]	3–13	28–60	34–80
Electric heating with coal [66]	35–50	19–40	21–39
Electric heating with scrap tires [5]	16–21	9–10	67–72
Electric heating with catalyst (zeolite,alumina) [67]	N.R[a]	N.R.	36–42
Carbon components in pyrolysis-oil	C5-C18		>C18
Microwave heating with particulate-carbon [65]	87 wt%		7 wt% (C19-C30)
Electric heating by jacketed electric heater [22]	19 wt%		81 wt% (C19-C35)
Electric heating by electric furnace [10][b]	45 wt%		55 wt% (C19-C29)
Electric heating by electric oven [5][b]	65 wt%		35 wt% (C19-C28)

[a] *N.R.—not reported;* [b] *The data from literature were estimated from the boiling point curve of waste oil obtained in their waste oil pyrolysis studies.*

Lam et al. [65] claimed that the different product compositions can be attributed to the use of the microwave-heated carbon bed in the experimental set-up, and the different heat distributions present during microwave pyrolysis. In microwave pyrolysis, the applied microwave radiation is targeted to and heats mainly the microwave-receptive particulate-carbon, creating a localized reaction 'hot zone' in which the added waste oil becomes totally immersed, providing excellent heat transfer and cracking capacity to crack the waste oil (C11-C40 hydrocarbons) to mainly C5-C18 hydrocarbons that can be re-condensed into pyrolysis-oil. The C5-C18 hydrocarbons, together with other cracked hydrocarbons, then vaporise as pyrolysis-volatiles, leave the hot carbon bed, and move into the vapour zone (the space above the carbon bed in the reactor) before being driven out of the reactor by the N_2 purge gas. The authors suggested that the pyrolysis-volatiles in the vapour zone were less likely to undergo further secondary reactions (e.g., secondary thermal cracking, carbonization) to form incondensable pyrolysis-gases and chars as there may not be enough thermal energy to supply the endothermic enthalpy to drive the secondary reactions. This is different than in the case of conventional electric heating from which the thermal energy is externally applied to the reactor and heats all the substances in the reactor including the evolved pyrolysis-volatiles, the surrounding gases, and the reactor chamber itself. This suggests that the pyrolysis-volatiles in conventional heating, being in contact with the hot surrounding gases and the walls of the reactor chamber, are likely to encounter a reaction 'hot zone' where more thermal energy is present in order for secondary reactions to occur. As a result, the pyrolysis-volatiles under these conditions are likely to undergo more secondary reactions than occur during microwave heating. In the former process, the hydrocarbon components in the pyrolysis-volatiles are further cracked to incondensable, lighter hydrocarbons (C1-C4 hydrocarbons) via secondary thermal cracking reactions, some of which are further transformed into chars via tertiary cracking or carbonization reactions. This leads to higher yields of both pyrolysis-gases and char residues and lower yield of pyrolysis-oil. Similar differences between conventional and microwave pyrolysis have also been observed during the treatment of other types of waste [62,68]. Other possible explanations that have been proposed to account for this difference include the microwave heating process itself, which has been shown to produce different products from conventional heating when all other factors are held equal [59,62], and the creation of free elections on the surface of the carbon particles as a result of microwave-induction, which may influence the reaction pathway [7].

12.3.3 ENERGY RECOVERY

Vigorous efforts have been made to estimate the energy recovery in microwave pyrolysis of waste materials [65,69–72]. The estimate provides a useful measure of the energy efficiency of the process, which is an important factor that determines the viability of this type of pyrolysis process, especially in scaling and optimising the design and operation to the commercial level. While this information has been revealed in microwave pyrolysis studies of several wastes, it was found that the existing literature is limited to pyrolysis performed in batch or semi-batch operation in which the feedstocks were added initially in one batch before being subjected to pyrolysis. Limited information is available concerning the energy balance in the pyrolysis that is performed in continuous operation.

It was revealed from these studies that microwave pyrolysis process is capable of recovering pyrolysis products (e.g., hydrocarbon oils) whose calorific value are many times greater than the amount of electrical energy used in the operation of the process, showing both a positive energy ratio (energy content of hydrocarbon products/electrical energy supplied for microwave heating) and a high energy output. For example, Lam et al. [65] have recently demonstrated through their studies on microwave pyrolysis of waste oil that the pyrolysis can be performed in a continuous operation, and the pyrolysis apparatus described is capable of treating waste oil with a positive energy ratio of 8 and a net energy output of 179 MJ/h. In particular, the oil product showed significantly high recovery (~90%) of the energy content of the waste oil. The authors also suggested that the favourable situation would be even more apparent during the operation of pilot or industrial scale equipment in which attempts to improve heat integration and recovery have been implemented.

However, it should be noted that the high energy recovery ratios observed in these pyrolysis studies involve the assumption that the only energy input of the process is the electrical energy used in the pyrolysis reactor. In practice lower energy ratios would be realised in which additional energy inputs have been taken into account, including the energy needed for the collection and transport of waste materials to the processing plant, and for the refining of the pyrolysis products (e.g., hydrocarbon oils) if they need to be further processed to produce a gas or liquid fuel.

Overall, it was established that microwave pyrolysis process show high recovery (60%–80%) of the energy input to the system (e.g., electrical energy input plus the calorific value of the added waste material) [65,69,70,72]. The recovered energy in the form of oil and gaseous products can be potentially used as a fuel source, e.g., on-site generation of electrical energy to power the microwave pyrolysis system. Furthermore, inclusion of heat integration and recovery systems to recover energy loss from the pyrolysis reactor (e.g., insulating the reaction vessel) would further increase the amount of energy that can be recovered from the system. The review on the existing literature indicates that the microwave pyrolysis method may be an energetically viable means of recycling waste materials into useful pyrolysis products, in addition to a disposal method for the waste.

12.4 APPLICATION OF MICROWAVE PYROLYSIS IN WASTE TO ENERGY PROCESSES

In view of the many advantages shown by microwave pyrolysis, studies have been performed to investigate the possible development of microwave pyrolysis process for efficient treatment and recovery of potentially valuable hydrocarbon feedstocks from waste materials. The distinct advantages shown by microwave pyrolysis (see Section 3.2) may lead to the potential for the greater production of desirable pyrolysis products, such as gaseous hydrocarbons and liquid hydrocarbon oils that can potentially be used as a fuel or chemical feedstock, while at the same time serving the purpose of recycling and disposing of waste materials. In view of that, the aim of the pyrolysis process would be to obtain valuable products that can be further used in other chemical processes and should demonstrate a commercial opportunity rather than a problem for the disposal of waste material. Thus, research on microwave pyrolysis processes of waste materials has always concentrated effort on clarifying the variation of product spectra with the nature of the load and the process conditions.

So far there has not been a consensus on what are the main products produced during the microwave pyrolysis of waste materials. In fact, there have been many explanations for the differences observed in the experiments conducted by different researchers, although some researchers have managed to obtain somewhat similar product yields. In general, three classes of product are obtained from the pyrolysis, that is: gases, oils, and char. Table 2 shows an example of the different yields of products obtained by different sets of researchers investigating energy recovery from microwave pyrolysis of waste materials.

TABLE 2: Product yield (wt%) from microwave pyrolysis of waste materials for energy recovery.

Research	Gases	Oil	Char
Waste automotive engine oil [65]	8	85	7
Plastic waste [7]	19–21	79–81	0
Sewage sludge [73,74]	36–63	2–8	30–60
Used car tyre [75]	10	50	40

In addition to the differences in product yield, varied product spectra were also found in the different studies. Domínguez et al. [74] and Zuo et al. [73] obtained high yields of syngas (H_2+CO) and low yields of CO_2 and CH_4 in their studies on microwave pyrolysis of sewage sludge. In contrast, Lam et al. [65,71] reported high yields of condensable oil product containing substantial concentrations of light aliphatic and aromatic hydrocarbons in their microwave pyrolysis studies of waste oil, but with a comparatively lower production of gaseous product that contains light aliphatic hydrocarbons and syngas. The similarities and differences in these studies were postulated to be mainly due to the influence of operating or experimental conditions (i.e., temperature, heating source, and pressure) and the nature of the load (i.e., chemical composition and source).

However, toxic metals were surprisingly found to be present in the oil fractions following condensation of the pyrolysis products from microwave pyrolysis of certain wastes such as waste oil [76] and oil-contaminated drill cuttings [33]. According to the authors, metals (e.g., Pb, Fe, Cu and Ni) are mainly present in the oils as metallic compounds; these compounds would turn into volatiles during pyrolysis processes at a reaction temperature of 600 °C and above. In addition, some metals (e.g., V, Cd) are condensed (accumulated) on the particulate matter produced during the process [76]. As a result, the metals condensed on the particulate matter and the metallic compounds (depending on their volatility) are likely to escape from the pyrolysis chamber with the other gaseous products during the pyrolysis process, causing the formation of undesired oil products containing toxic metals. As a result, efforts are being made to incorporate additional processes such as hot gas cleaning and previous demetalisation of waste materials in order to obtain oil products that are free of metals.

Toxic semi-volatile PAH compounds such as naphthalene, acenaphthylene, phenanthrene, anthracene, and pyrene were also detected in the pyrolysis products from microwave pyrolysis of waste oil [65,76] and sewage sludge [77]. Thus, these studies imply that further investigation is needed to perfect the pyrolysis process conditions in order to improve the production of valuable pyrolysis products whilst controlling the formation of potentially toxic compounds described above. Some of the applications of microwave pyrolysis in the treatment and energy recovery from waste materials are further presented and discussed in the following Sections (4.1–4.5).

12.4.1 WASTE AUTOMOTIVE ENGINE OIL

Microwave pyrolysis has recently shown excellent potential for recycling waste oil [65,71,76,78]. By pyrolysing the waste oil in a modified microwave oven in the presence of a bed of particulate-carbon as the microwave-absorbent, hydrocarbons (of smaller molecular size than those present in the waste oil) and H_2 are generated and these have potential for use as either an energy source or industrial feedstock [65,78]. The waste oil is a poor microwave-absorbent on its own due to its non-polar nature, therefore it requires heating by contact with materials with high microwave absorbency in order to achieve pyrolytic thermal cracking [65].

The pyrolysis generated a high yield (88 wt.%) of a condensable oil product that contains substantial concentrations of potentially valuable light aliphatic and aromatic hydrocarbons [71], and with fuel properties comparable to transport-grade fuels [65]. The oil product showed a high recovery of the calorific value present in the waste oil, is relatively contaminant free with a low content of sulphur, oxygen, and residue, and is almost entirely free of metals [76]. In addition, the oil product is reported to show a low toxic risk and contain negligible or minor amounts of toxic polycyclic aromatic hydrocarbons (PAHs) compounds [76], which was the case for oil products from conventional electric-heated pyrolysis of waste oil [26,79]. The authors claimed that the oil product could potentially be treated and upgraded to transport grade fuels, or added to petroleum refinery as a chemical feedstock for further processing, although further studies are needed to confirm these possibilities. Figure 3 shows a schematic representation of the generation of oil product comparable to commercial gasoline fuel from microwave pyrolysis of the waste oil.

The pyrolysis process also produced an incondensable gaseous product that contains light aliphatic hydrocarbons and syngas that could potentially be used as a gaseous fuel and chemical feedstock [78]. In particular, the hydrogen obtained can be potentially used as a second-generation fuel or synthesis chemical, and the CO could be steam-reformed to produce more hydrogen. The light C2-C6 hydrocarbons generated can also be reformed easily to produce additional hydrogen or to be extracted for use as chemical feedstocks. The gaseous product could be burned directly in gas engine or fuel cells, or upgraded to produce hydrogen, and synthetic fuel (via Fischer–Tropsch process). Figure 4 shows a schematic representation of the generation of light aliphatic hydrocarbons and syngas from microwave pyrolysis of the waste oil.

Additionally, the pyrolysis generated a char product that contained the majority of metals originally present in the waste oil, providing a convenient opportunity for the efficient recovery of these metals [76]. The char can be readily separated from the particulate-carbon particles by sieving, and the particulate-carbon bed can be repeatedly re-used as the microwave-absorbent (heating medium) after the separation, as a result of the fact that the majority of the metals were found to be present within the char particulates, resulting in very low levels of residual metals being retained within the particulate-carbon bed itself. The microwave apparatus was operated with an electrical power input of 7.5 kW and was capable of processing waste oil at a flow rate of 5 kg/h with a positive energy ratio of 8 (energy content of hydrocarbon products/electrical energy supplied for microwave heating) and a net energy output of 179,390 kJ/h [65].

The product compositions, which are different to those formed in conventional pyrolysis of oils, can be attributed to the unique heating mode and the chemical environment present during microwave pyrolysis, and chemical mechanisms for the production of the various products were proposed [65,78]. The authors demonstrated that microwave pyrolysis offers an exciting new approach to treat and transform the waste oil into valuable hydrocarbon feedstocks and gases.

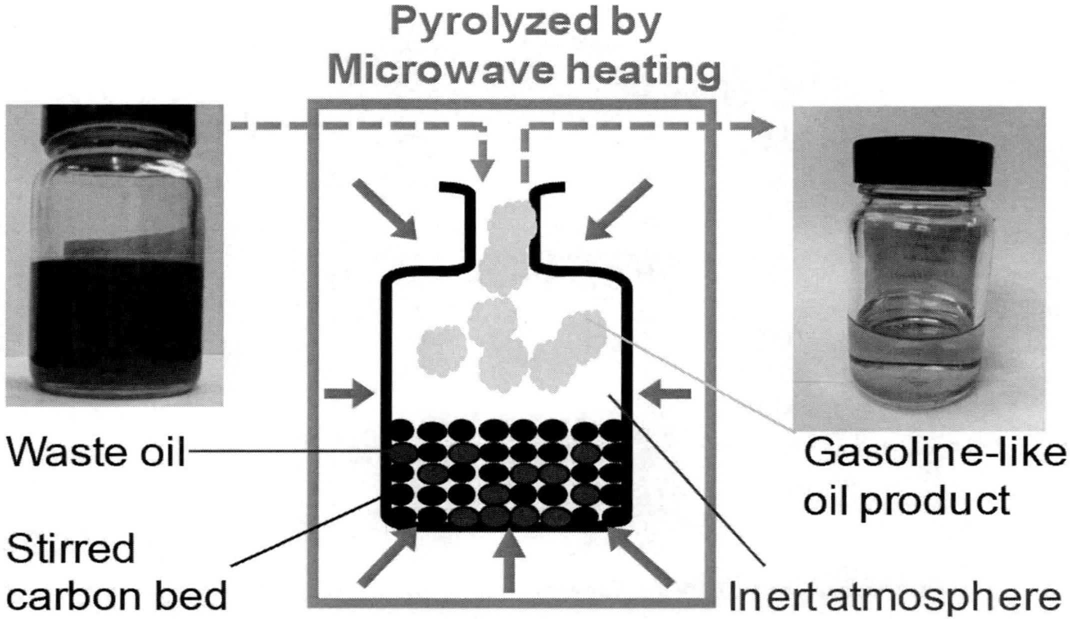

FIGURE 3. Schematic representation of the generation of gasoline-like oil product from microwave pyrolysis of waste oil [65].

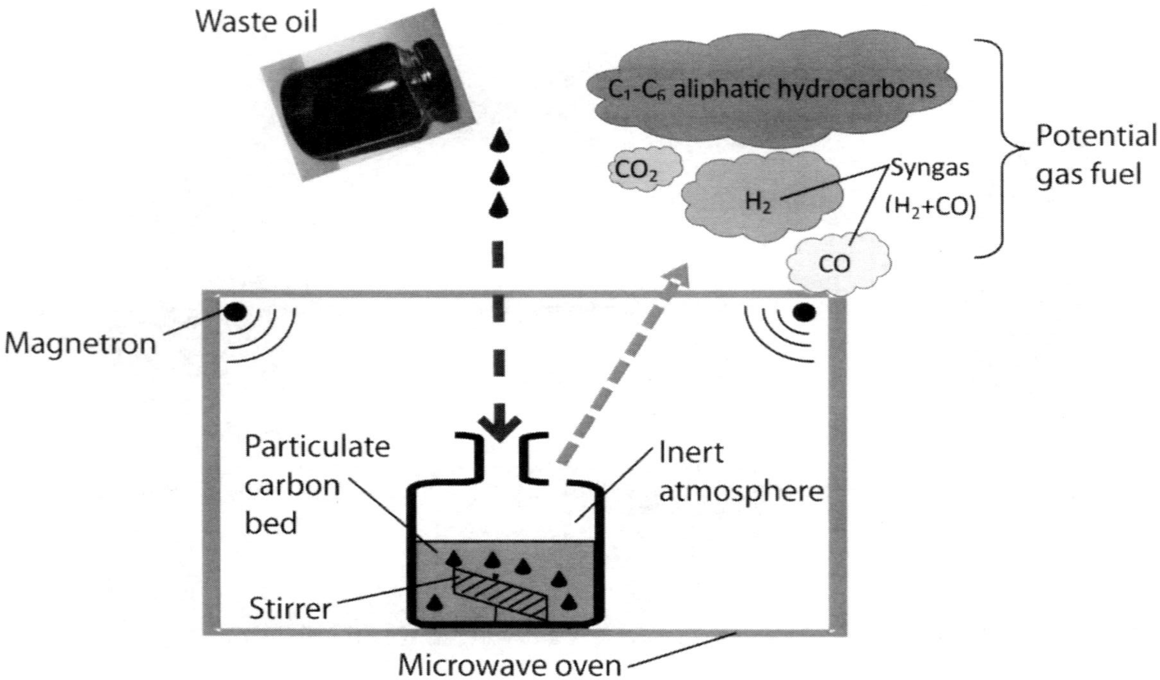

FIGURE 4: Schematic representation of the generation of light aliphatic hydrocarbons and syngas from microwave pyrolysis of waste oil [78].

12.4.2 PLASTIC WASTES

Microwave pyrolysis has been successfully applied to high density polyethylene using a stirred carbon bed, producing a high yield of oil/waxes (80%) at 500°C and 600°C [7]. This work has been extended to aluminium-coated packaging waste, where the pyrolysis process was performed to separate aluminium from aluminium-coated polymer laminates found in toothpaste tubes and de-pulped drink-cartons (e.g., Tetrapacks) [7]. The process showed great promise as a separation technology as almost the entire aluminium fraction could be recovered from the waste packaging.

12.4.3 BIOMASS WASTES

Sewage sludge is an environmentally hazardous, high-volume waste that has become a major concern for modern society. The existing disposal processes, such as landfilling and incineration, are becoming increasingly impracticable as concerns over environmental pollution and high treatment costs are recognised due to contaminants present in sewage sludge [35]. In addition, the high treatment costs are partly due to the need to remove the high amounts of water present in the sludge. In view of the limitation associated with current disposal methods, a great deal of studies has been performed to develop microwave pyrolysis as an alternative to treat and recycle sewage sludge [62,74,77,80,81]. In these studies, sewage sludge was mixed with small amounts of the char produced from previous runs and subjected to microwave pyrolysis treatment at 1040°C. The authors demonstrated that the use of microwave heating is effective in drying the sewage sludge during the initial stage of the pyrolysis process. Then, the steam produced by the microwave-induced drying could be used to gasify the products generated from the pyrolysis process. The authors claimed that the microwave pyrolysis method showed advantages over conventional pyrolysis in providing drying, pyrolysis, and gasification treatments to sewage sludge at the same time while generating a high yield of a gaseous product containing valuable syngas (up to 66 vol%) and a small amount of oil product with a low content of PAHs. In contrast, they reported that the oil product obtained from conventional pyrolysis of sewage sludge contained mostly of PAHs. Tian et al. [72] have recently demonstrated the use of microwave pyrolysis to produce bio-oil from sewage sludge. The authors claimed that a high yield of bio-oil could be obtained at a microwave power ranging from 400 to 600 W, and the bio-oil could potentially be used as a bio-fuel.

In addition to sewage sludge, microwave pyrolysis has been used to produce hydrogen rich fuel gas from coffee bean hulls [12], rice straw [69], straw bale [70]. The authors claimed that a high yield of gaseous product (~70 wt%) was obtained at a pyrolysis temperature of 1000 °C. The gaseous product also showed a greater content of H_2 (35–50 vol%) and syngas (50–72 vol%) than that produced during conventional electric-heated pyrolysis under similar conditions (with a H_2 content of 30 vol% and a syngas content of 53 vol%).

Microwave pyrolysis has also been employed to convert wood block to tar (up to 30 wt%), oil, and charcoal [82]. The diffuse nature of the electromagnetic field allows microwave heating to evenly heat many substances in bulk [61], and thus allowing the processing of large wood samples without the need of a pre-treatment process such as pulverisation. The author demonstrated that microwave pyrolysis showed advantages in providing a rapid heating and less power consumption when compared with conventional pyrolysis.

12.4.4 HAZARDOUS WASTE PROCESSING

Microwave pyrolysis has also been successfully used to transform car tyres into carbon black, steel, liquid hydrocarbon oil, and gaseous hydrocarbons [75]. In addition, the pyrolysis has been applied to decontaminate oil-contaminated drill cuttings [33]. The authors demonstrated that the pyrolysis can be performed without the need for an appropriate intermediate microwave receptor, as the water present in the cuttings can directly be used as a microwave-absorbent to generate heat in order for pyrolysis to occur. However, the extent to which the process actually decontaminates the cuttings has yet to be fully investigated. Chlorodifluoromethane is another hazardous waste that has been processed using microwave pyrolysis in a microwave-heated fluidised bed [83]. The authors claimed that microwave pyrolysis showed advantages in providing a rapid heating process to pyrolyse the waste without overheating the reactor walls.

12.4.5 PRODUCTION OF SYNTHETIC FUEL

Coal has been processed with microwave pyrolysis at high temperatures (up to 1300°C) using a variety of intermediate microwave receptors (i.e., coke, Fe_3O_4, CuO) in order to produce condensable tars suitable for use as a fuel [39]. A 20 wt% yield of condensable tar was achieved in the presence of coke, whereas a yield of 27 wt% was recorded in the presence of Fe_3O_4, and a maximum yield of 49 wt% was obtained in the presence of CuO. The authors claimed that coal (which is known to be a poor microwave-absorbent) can be efficiently processed with microwave pyrolysis in the presence of inorganic oxides as the microwave receptor, although the coal heats up slowly on its own when subjected to microwave radiation and is transformed into carbon black.

Oil shale is another material that has been processed by microwave pyrolysis at a temperature of 700 °C, producing 6 wt% of oil product and 10 wt% of gaseous product [41]. The authors found that the product yields obtained by microwave pyrolysis were similar to those obtained by conventional pyrolysis, but they claimed that microwave pyrolysis promoted a different product composition as more cracking was observed and this resulted in production of lower amounts of polar, sulphur, and nitrogen compounds.

12.5 CONCLUSIONS

Up to now research in microwave pyrolysis has been centred on its application to treating wastes such as plastic waste [7], sewage sludge [74], scrap tyres [18], wood blocks [82], oil shales [41], and various organic wastes [84]. However, despite the variety of research that has been conducted on microwave pyrolysis, the growth of industrial microwave heating applications is hampered by an apparent lack of the understanding of microwave systems and the technical information for designing commercial equipment for these pyrolysis processes. Also, there have been no reports on the economic assessment of the microwave pyrolysis in order to determine the economic viability of the process, and limited information is available on the economic evaluation of the other pyrolysis processes in the literature, thus no economic comparisons have been made to date.

Consequently, despite the advantages shown by microwave pyrolysis, the use of this technology for waste treatment has not been extensively exploited. It was revealed that many important characteristics of the microwave pyrolysis process have yet to be raised or fully investigated, e.g., the influence of key process parameters on the yield and chemical composition of the product, the heating characteristic of a substance, and the production of any toxic compounds during the pyrolysis by microwave radiation. In addition, research on microwave pyrolysis has so far focused on pyrolysis performed in batch or semi-batch operation using typical process configurations that have also been used for conventional pyrolysis such as rotary kilns and fixed beds, where certain limitations of these processes have caused problems in treating real wastes such as waste oil. Thus, these limitations make it difficult for the pyrolysis processes to be optimized and demonstrated to be commercially attractive.

So far, there has been little research reported on the microwave pyrolysis of waste materials. It was established that current microwave pyrolysis techniques offer a number of advantages and show excellent potential for treating waste materials. However, it was found that such practice possesses limitations and uncertainties and there are still gaps to be filled in order to fully exploit the advantages of using microwave pyrolysis process in the treatment of waste materials. Hence, the main aim of using microwave pyrolysis is to provide an alternative pyrolysis process by making use of the high temperatures that the carbon bed can reach when subjected to a microwave field. This alternative way of heating is reported to have advantages over other conventional pyrolysis processes on account of better heat transfer to the waste materials, good control over the heating process as well as offering a very reducing chemical environment.

In summary, this literature review has revealed several aspects of microwave pyrolysis that need to be examined to investigate whether this particular pyrolysis technique is optimal for the treatment and energy recovery from waste materials. However, in view of mainly positive findings reported in the literature on the microwave pyrolysis studies, it would be worthwhile to carry on researching further aspects of microwave pyrolysis of waste materials in order to explore the full potential of this process. The optimisation of this process and the subsequent scale-up to a commercial scale will depend on how well the parameters involved in this new process and their relationships are understood.

REFERENCES

1. Latimer, J.S.; Hoffman, E.J.; Hoffman, G.; Fasching, J.L.; Quinn, J.G. Sources of petroleum-hydrocarbons in urban runoff. Water Air Soil Pollut. 1990, 52, 1–21.
2. Cormier, S.A.; Lomnicki, S.; Backes, W.; Dellinger, B. Origin and health impacts of emissions of toxic by-products and fine particles from combustion and thermal treatment of hazardous wastes and materials. Environ. Health Perspect. 2006, 114, 810–817.
3. Rowat, S.C. Incinerator toxic emissions: A brief summary of human health effects with a note on regulatory control. Med. Hypotheses 1999, 52, 389–396.
4. Dry, M.E. High quality diesel via the Fischer–Tropsch process—A review. J. Chem. Technol. Biotechnol. 2002, 77, 43–50.
5. Uçar, S.; Karagöz, S.; Yanik, J.; Saglam, M.; Yuksel, M. Copyrolysis of scrap tires with waste lubricant oil. Fuel Process. Technol. 2005, 87, 53–58.
6. Scheirs, J.; Kaminsky, W. Feedstock Recycling and Pyrolysis of Waste Plastics: Converting Waste Plastics into Diesel and Other Fuels; Wiley: Chichester, UK, 2006; Volume 27, pp. 2–785.
7. Ludlow-Palafox, C.; Chase, H.A. Microwave induced pyrolysis of plastic wastes. Ind. Eng. Chem. Res. 2001, 40, 4749–4756.
8. Dry, M.E. The Fischer–Tropsch process: 1950–2000. Catal. Today 2002, 71, 227–241.
9. Song, G.-J.; Seo, Y.-C.; Pudasainee, D.; Kim, I.-T. Characteristics of gas and residues produced from electric arc pyrolysis of waste lubricating oil. Waste Manag. 2010, 30, 1230–1237.
10. Sinağ, A.; Gülbay, S.; Uskan, B.; Uçar, S.; Özgürler, S.B. Production and characterization of pyrolytic oils by pyrolysis of waste machinery oil. J. Hazard. Mater. 2010, 173, 420–426.
11. Onwudili, J.A.; Insura, N.; Williams, P.T. Composition of products from the pyrolysis of polyethylene and polystyrene in a closed batch reactor: Effects of temperature and residence time. J. Anal. Appl. Pyrolysis 2009, 86, 293–303.
12. Domínguez, A.; Menéndez, J.A.; Fernández, Y.; Pis, J.J.; Nabais, J.M.V.; Carrott, P.J.M.; Carrott, M.M.L.R. Conventional and microwave induced pyrolysis of coffee hulls for the production of a hydrogen rich fuel gas. J. Anal. Appl. Pyrolysis 2007, 79, 128–135.

13. Carlson, T.R.; Cheng, Y.-T.; Jae, J.; Huber, G.W. Production of green aromatics and olefins by catalytic fast pyrolysis of wood sawdust. Energy Environ. Sci. 2011, 4, 145–161.
14. Bridgwater, A.V. Review of fast pyrolysis of biomass and product upgrading. Biomass Bioenergy 2012, 38, 68–94.
15. Osepchuk, J.M. Microwave power applications. IEEE Trans. Microwave Theory Tech. 2002, 50, 975–985.
16. Marken, F.; Sur, U.K.; Coles, B.A.; Compton, R.G. Focused microwaves in electrochemical processes. Electrochim. Acta 2006, 51, 2195–2203.
17. Ramasamy, K.K.; T-Raissi, A. Hydrogen production from used lubricating oils. Catal. Today. 2007, 129, 365–371.
18. Ludlow-Palafox, C.; Chase, H.A. Microwave pyrolysis of plastic wastes. In Feedstock Recycling and Pyrolysis of Waste Plastics; Scheirs, J., Kaminsky, W., Eds.; John Wiley and Sons Ltd.: Chichester, UK, 2006; pp. 569–594.
19. Arpa, O.; Yumrutas, R.; Demirbas, A. Production of diesel-like fuel from waste engine oil by pyrolitic distillation. Appl. Energy 2010, 87, 122–127.
20. Fuentes, M.J.; Font, R.; Gómez-Rico, M.F.; Martín-Gullón, I. Pyrolysis and combustion of waste lubricant oil from diesel cars: Decomposition and pollutants. J. Anal. Appl. Pyrol. 2007, 79, 215–226.
21. Balat, M.; Demirbas, M.F. Pyrolysis of waste engine oil in the presence of wood ash. Energy Sources Part A 2009, 31, 1494–1499.
22. Kim, Y.S.; Jeong, S.U.; Yoon, W.L.; Yoon, H.K.; Kim, S.H. Tar-formation kinetics and adsorption characteristics of pyrolyzed waste lubricating oil. J. Anal. Appl. Pyrolysis 2003, 70, 19–33.
23. Kim, S.S.; Chun, B.H.; Kim, S.H. Non-isothermal pyrolysis of waste automobile lubricating oil in a stirred batch reactor. Chem. Eng. J. 2003, 93, 225–231.
24. Juniper Consultancy Services Ltd. Pyrolysis and Gasification of Wastes; Juniper Consultancy Services Ltd.: London, UK, 1997.
25. Williams, P.T.; Besler, S. Polycyclic aromatic hydrocarbons in waste derived pyrolytic oils. J. Anal. Appl. Pyrolysis 1994, 30, 17–33.
26. Domeño, C.; Nerín, C. Fate of polyaromatic hydrocarbons in the pyrolysis of industrial waste oils. J. Anal. Appl. Pyrolysis 2003, 67, 237–246.
27. Lázaro, M.J.; Moliner, R.; Suelves, I.; Nerín, C.; Domeño, C. Valuable products from mineral waste oils containing heavy metals. Environ. Sci. Technol. 2000, 34, 3205–3210.
28. Holland, K.M. Apparatus for Waste Pyrolysis. U.S. Patent 5,387,321, 7 February 1995.
29. Holland, K.M. Process of Destructive Distillation of Organic Material. U.S. Patent 5,330,623, 19 July 1994.
30. Bilali, L.; Benchanaa, M.; El harfi, K.; Mokhlisse, A.; Outzourhit, A. A detailed study of the microwave pyrolysis of the Moroccan (Youssoufia) rock phosphate. J. Anal. Appl. Pyrolysis 2005, 73, 1–15.
31. Li, H.; Liao, L.; Liu, L. Kinetic Investigation into the Non-thermal microwave effect on the ring-opening polymerization of ε-Caprolactone. Macromol. Rapid Commun. 2007, 28, 411–416.

32. Meredith, R. Engineers' Handbook of Industrial Microwave Heating; The Institution of Electrical Engineers: London, UK, 1998.
33. Robinson, J.; Snape, C.; Kingman, S.; Shang, H. Thermal desorption and pyrolysis of oil contaminated drill cuttings by microwave heating. J. Anal. Appl. Pyrolysis 2008, 81, 27–32.
34. Metaxas, A.C. Foundations of Electroheat; John Wiley and Sons Ltd.: Chichester, UK, 1996.
35. Menéndez, J.A.; Arenillas, A.; Fidalgo, B.; Fernández, Y.; Zubizarreta, L.; Calvo, E.G.; Bermúdez, J.M. Microwave heating processes involving carbon materials. Fuel Process. Technol. 2010, 91, 1–8.
36. Zhang, X.; Hayward, D.O. Applications of microwave dielectric heating in environment-related heterogeneous gas-phase catalytic systems. Inorg. Chim. Acta 2006, 359, 3421–3433.
37. Udalov, E.; Bolotov, V.; Tanashev, Y.; Chernousov, Y.; Parmon, V. Pyrolysis of liquid hexadecane with selective microwave heating of the catalyst. Theor. Exp. Chem. 2011, 46, 384–392.
38. Menéndez, J.A.; Domínguez, A.; Fernández, Y.; Pis, J.J. Evidence of Self-Gasification during the Microwave-Induced Pyrolysis of Coffee Hulls. Energy Fuel. 2007, 21, 373–378.
39. Monsef-Mirzai, P.; Ravindran, M.; McWhinnie, W.R.; Burchill, P. Rapid microwave pyrolysis of coal: Methodology and examination of the residual and volatile phases. Fuel 1995, 74, 20–27.
40. Monsef-Mirzai, P.; Ravindran, M.; McWhinnie, W.R.; Burchil, P. The use of microwave heating for the pyrolysis of coal via inorganic receptors of microwave energy. Fuel 1992, 71, 716–717.
41. El harfi, K.; Mokhlisse, A.; Chanaa, M.B.; Outzourhit, A. Pyrolysis of the Moroccan (Tarfaya) oil shales under microwave irradiation. Fuel 2000, 79, 733–742.
42. Fernández, Y.; Arenillas, A.; Díez, M.A.; Pis, J.J.; Menéndez, J.A. Pyrolysis of glycerol over activated carbons for syngas production. J. Anal. Appl. Pyrolysis 2009, 84, 145–150.
43. Menéndez, J.A.; Inguanzo, M.; Pis, J.J. Microwave-induced pyrolysis of sewage sludge. Water Res. 2002, 36, 3261–3264.
44. Menéndez, J.A.; Menéndez, E.M.; Garcia, A.; Parra, J.B.; Pis, J.J. Thermal treatment of active carbons: A comparison between microwave and electrical heating. Int. Microwave Power Inst. 1999, 34, 137–143.
45. Liu, C.C.; Walters, A.; Vannice, M. Measurement of electrical properties of a carbon black. Carbon 1995, 33, 1699–1708.
46. Fidalgo, B.; Arenillas, A.; MenÈndez, J.A. Influence of porosity and surface groups on the catalytic activity of carbon materials for the microwave-assisted CO2 reforming of CH4. Fuel 2010, 89, 4002–4007.
47. Roberts, B.A.; Strauss, C.R. Toward rapid, "green", predictable microwave-assisted synthesis. Acc. Chem. Res. 2005, 38, 653–661.
48. Nuchter, M.; Ondruschka, B.; Bonrath, W.; Gum, A. Microwave assisted synthesis—A critical technology overview. Green Chem. 2004, 6, 128–141.
49. Strauss, C.; Rooney, D. Accounting for clean, fast and high yielding reactions under microwave conditions. Green Chem. 2010, 12, 1340–1344.

50. Fantini, M.; Zuliani, V.; Spotti, M.A.; Rivara, M. Microwave assisted efficient synthesis of imidazole-based privileged structures. J. Comb. Chem. 2009, 12, 181–185.
51. De Paolis, O.; Teixeira, L.; Török, B. Synthesis of quinolines by a solid acid-catalyzed microwave-assisted domino cyclization-aromatization approach. Tetrahedron Lett. 2009, 50, 2939–2942.
52. Herrero, M.A.; Kremsner, J.M.; Kappe, C.O. Nonthermal microwave effects revisited: On the importance of internal temperature monitoring and agitation in microwave chemistry. J. Org. Chem. 2007, 73, 36–47.
53. Hosseini, M.; Stiasni, N.; Barbieri, V.; Kappe, C.O. Microwave-assisted asymmetric organocatalysis. A probe for nonthermal microwave effects and the concept of simultaneous cooling. J. Org. Chem. 2007, 72, 1417–1424.
54. De la Hoz, A.; Diaz-Ortiz, A.; Moreno, A. Microwaves in organic synthesis. Thermal and non-thermal microwave effects. Chem. Soc. Rev. 2005, 34, 164–178.
55. Schmink, J.R.; Leadbeater, N.E. Probing "microwave effects" using Raman spectroscopy. Org. Biomol. Chem.2009, 7, 3842–3846.
56. Razzaq, T.; Kremsner, J.M.; Kappe, C.O. Investigating the existence of nonthermal/specific microwave effects using silicon carbide heating elements as power modulators. J. Org. Chem. 2008, 73, 6321–6329.
57. Bagnell, L.; Cablewski, T.; Strauss, C.R.; Trainor, R.W. Reactions of allyl phenyl ether in high-temperature water with conventional and microwave heating. J. Org. Chem. 1996, 61, 7355–7359.
58. Jones, D.A.; Lelyveld, T.P.; Mavrofidis, S.D.; Kingman, S.W.; Miles, N.J. Microwave heating applications in environmental engineering—A review. Resour. Conserv. Recycl. 2002, 34, 75–90.
59. Fernández, Y.; Arenillas, A.; Bermúdez, J.M.; Menéndez, J.A. Comparative study of conventional and microwave-assisted pyrolysis, steam and dry reforming of glycerol for syngas production, using a carbonaceous catalyst. J. Anal. Appl. Pyrolysis 2010, 88, 155–159.
60. Appleton, T.J.; Colder, R.I.; Kingman, S.W.; Lowndes, I.S.; Read, A.G. Microwave technology for energy-efficient processing of waste. Appl. Energy 2005, 81, 85–113.
61. Lei, H.; Ren, S.; Julson, J. The effects of reaction temperature and time and particle size of corn stover on microwave pyrolysis. Energy Fuel. 2009, 23, 3254–3261.
62. Menéndez, J.A.; Domínguez, A.; Inguanzo, M.; Pis, J.J. Microwave pyrolysis of sewage sludge: Analysis of the gas fraction. J. Anal. Appl. Pyrolysis 2004, 71, 657–667.
63. Committee on Microwave Processing of Materials: An Emerging Industrial Technology, National Materials Advisory Board, Commission on Engineering and Technical Systems and National Research Council (US). Microwave Processing of Materials; National Academy Press: Washington, DC, USA, 1994; Volume 473.
64. Du, Z.; Li, Y.; Wang, X.; Wan, Y.; Chen, Q.; Wang, C.; Lin, X.; Liu, Y.; Chen, P.; Ruan, R. Microwave-assisted pyrolysis of microalgae for biofuel production. Bioresour. Technol. 2011, 102, 4890–4896.
65. Lam, S.S.; Russell, A.D.; Lee, C.L.; Chase, H.A. Microwave-heated pyrolysis of waste automotive engine oil: Influence of operation parameters on the yield, composition, and fuel properties of pyrolysis oil. Fuel 2012, 92, 327–339.

66. Lázaro, M.J.; Moliner, R.; Suelves, I.; Domeño, C.; Nerín, C. Co-pyrolysis of a mineral waste oil/coal slurry in a continuous-mode fluidized bed reactor. J. Anal. Appl. Pyrolysis 2002, 65, 239–252.
67. Demirbas, A. Gasoline-like fuel from waste engine oil via catalytic pyrolysis. Energy Sources Part A 2008, 30, 1433–1441.
68. Domínguez, A.; Menéndez, J.A.; Inguanzo, M.; Pis, J.J. Investigations into the characteristics of oils produced from microwave pyrolysis of sewage sludge. Fuel Process. Technol. 2005, 86, 1007–1020.
69. Huang, Y.F.; Kuan, W.H.; Lo, S.L.; Lin, C.F. Hydrogen-rich fuel gas from rice straw via microwave-induced pyrolysis. Bioresour.Technol. 2010, 101, 1968–1973.
70. Zhao, X.; Zhang, J.; Song, Z.; Liu, H.; Li, L.; Ma, C. Microwave pyrolysis of straw bale and energy balance analysis. J. Anal. Appl. Pyrolysis 2011, 92, 43–49.
71. Lam, S.S.; Russell, A.D.; Chase, H.A. Microwave pyrolysis, a novel process for recycling waste automotive engine oil. Energy 2010, 35, 2985–2991.
72. Tian, Y.; Zuo, W.; Ren, Z.; Chen, D. Estimation of a novel method to produce bio-oil from sewage sludge by microwave pyrolysis with the consideration of efficiency and safety. Bioresour. Technol. 2011, 102, 2053–2061.
73. Zuo, W.; Tian, Y.; Ren, N. The important role of microwave receptors in bio-fuel production by microwave-induced pyrolysis of sewage sludge. Waste Manag. 2011, 31, 1321–1326.
74. Domínguez, A.; Fernández, Y.; Fidalgo, B.; Pis, J.J.; Menéndez, J.A. Bio-syngas production with low concentrations of CO2 and CH4 from microwave-induced pyrolysis of wet and dried sewage sludge. Chemosphere 2008, 70, 397–403.
75. Yatsun, A.; Konovalov, P.; Konovalov, N. Gaseous products of microwave pyrolysis of scrap tires. Solid Fuel Chem. 2008, 42, 187–191.
76. Lam, S.S.; Russell, A.D.; Chase, H.A. Pyrolysis using microwave heating: A sustainable process for recycling used car engine oil. Ind. Eng. Chem. Res. 2010, 49, 10845–10851.
77. Domínguez, A.; Menéndez, J.A.; Inguanzo, M.; Pís, J.J. Production of bio-fuels by high temperature pyrolysis of sewage sludge using conventional and microwave heating. Bioresour. Technol. 2006, 97, 1185–1193.
78. Lam, S.S.; Russell, A.D.; Lee, C.L.; Lam, S.K.; Chase, H.A. Production of hydrogen and light hydrocarbons as a potential gaseous fuel from microwave-heated pyrolysis of waste automotive engine oil. Int. J. Hydrog. Energy 2012, 37, 5011–5021.
79. Nerin, C.; Domeno, C. Determination of polyaromatic hydrocarbons and some related compounds in industrial waste oils by GPC-HPLC-UV. Analyst 1999, 124, 67–70.
80. Menendez, J.A.; Dominguez, A.; Inguanzo, M.; Pis, J.J. Microwave-induced drying, pyrolysis and gasification (MWDPG) of sewage sludge: Vitrification of the solid residue. J. Anal. Appl. Pyrolysis 2005, 74, 406–412.
81. Domínguez, A.; Menéndez, J.A.; Inguanzo, M.; Bernad, P.L.; Pis, J.J. Gas chromatographic-mass spectrometric study of the oil fractions produced by microwave-assisted pyrolysis of different sewage sludges. J. Chromatogr. A 2003, 1012, 193–206.
82. Miura, M.; Kaga, H.; Sakurai, A.; Kakuchi, T.; Takahashi, K. Rapid pyrolysis of wood block by microwave heating. J. Anal. Appl. Pyrolysis 2004, 71, 187–199.

83. Kim, H.C.; Kim, H.Y.; Woo, S.I. Fast pyrolysis of chlorodifluoromethane in a microwave-heated fluidized bed. J. Chem. Eng. Jpn. 1999, 32, 171–176.
84. Chemat, F.; Poux, M. Microwave assisted pyrolysis of urea supported on graphite under solvent-free conditions. Tetrahedron Lett. 2001, 42, 3693–3695.

PART IV

HYDROTHERMAL CARBONIZATION

CHAPTER 13

Hydrothermal Upgrading of Korean MSW for Solid Fuel Production: Effect of MSW Composition

DAEGI KIM, PANDJI PRAWISUDHA, AND KUNIO YOSHIKAWA

13.1 INTRODUCTION

In recent years, the global issue in the energy field is that with the combination of increasing energy consumption and the steady depletion of fossil fuel reserves, coal can only be used to last 122 years on the basis of the 2008 production rate. This, together with the global environmental issues of the appropriate treatment of increasing municipal solid waste (MSW) has prompted a global research to develop alternative energy resources as well as to reduce CO_2 emissions by using renewable energy from biomass and waste [1–3]. Korean government has had an interest to employ a new MSW treatment system, namely, the mechanical biological treatment (MBT) system. The MBT system concepts for waste processing evolved in Germany and incorporated two stages of mechanical treatment (MT) and biological treatment (BT). The bigger size MSW separated by the mechanical treatment (MT) as combustible matter is processed to RDF (refuse derived fuel) for energy generation, while the separated MSW after MT (MT residue) is used for producing organic fertilizer and biogas (CH_4) by employing the BT stage. The MBT system enables us to circulate resources and to reduce the greenhouse gas emissions, while getting the profit by making renewable resource fuels from MSW [2] to reduce the quantity of waste sent to landfill and to increase the potential recovery of resources. This system acts as a pretreatment system for the next step of processing [3, 4].

However, in the MBT system, the BT stage has common problems requiring long treatment time, more than 1 week to 1 month, with unpleasant smells [5, 6]. Especially, food residues in Korea are inappropriate for composting due to the high salinity from food residue such as kimchi, which are fermented and stored in highly saline water [5].

The hydrothermal treatment is one of the thermochemical processes, treating waste in high-temperature and high-pressure water media to upgrade the material in a short time [7–10]. That is one of progressive technologies for converting MSW and biomass into useful energy resources because it can improve the dehydration and drying performances of high moisture content biomass as well as upgrade the property of the fuel produced from MSW.

MSW differs in quality and quantity depending on the policy and culture of the nation. The composition of MSW should be different according to seasons and sectors and affected by custom, living style, and so forth as well as regulation of the country. Separated MSW by MT in Korea was usually consisted of high food residue (40–50%), paper (30–40%), plastic, and so forth. Therefore, it has high moisture content of about 50–60% since the food residue affects the moisture content and major properties of MSW. Especially, Korean food residue has had high moisture and high salinity to affect the treatment solution. Its property should provide a negative product to treat MT residue in BT stage. If we want to utilize the MT residue as a solid fuel together with RDF the MT residue should be dehydrated, dried, upgraded, and compacted.

In this research, a mild reaction condition of subcritical water (180°C < T < 220°C, 1.8 MPa < P < 2.4 MPa) is employed in the hydrothermal treatment, focused on its effects on the surrogate MT residue (kimchi, paper, and their mixture). The aim of the present work is to demonstrate the improvement of the dehydration and drying performances and upgrading of solid products of surrogate MT residue in a short time and to make uniform solid fuel using the hydrothermal treatment.

13.2 OPERATION PRINCIPLE OF THE HYDROTHERMAL TREATMENT

The hydrothermal treatment employed in this research is utilized using high-temperature water to treat the raw material. Solid wastes are fed into the reactor, and then, about 200°C, 2 MPa saturated steam is supplied into the reactor for about 30 minutes and the blades are installed inside the reactor to mix the wastes for about 30–90 minutes. Then steam inside of the reactor will be discharged, condensed, and treated to be utilized as the boiler feed water again. The product is powder-like substance, and the moisture content is higher than the raw material, but it shows much improved dehydration and drying performances than the raw material. The hydrothermal reactions include the hydration, hydrous pyrolysis and decarboxylation, and so on, and the reaction temperature affects the properties of hydrothermal products [8–12]. The hydrothermal treatment has benefits of no requirement of pretreatment, no emissions and lower treatment temperature compared with other thermal treatment methods such as carbonization, pyrolysis, gasification, and incineration. Moreover, the hydrothermal treatment is not only reducing the recovery costs but also environmental friendly with no usage of chemical agents [13–16]. Recently, the hydrothermal treatment has attracted interests as a possible application producing coal-like solid fuel from MSW and biomass resources with high moisture and oxygen contents [8–22].

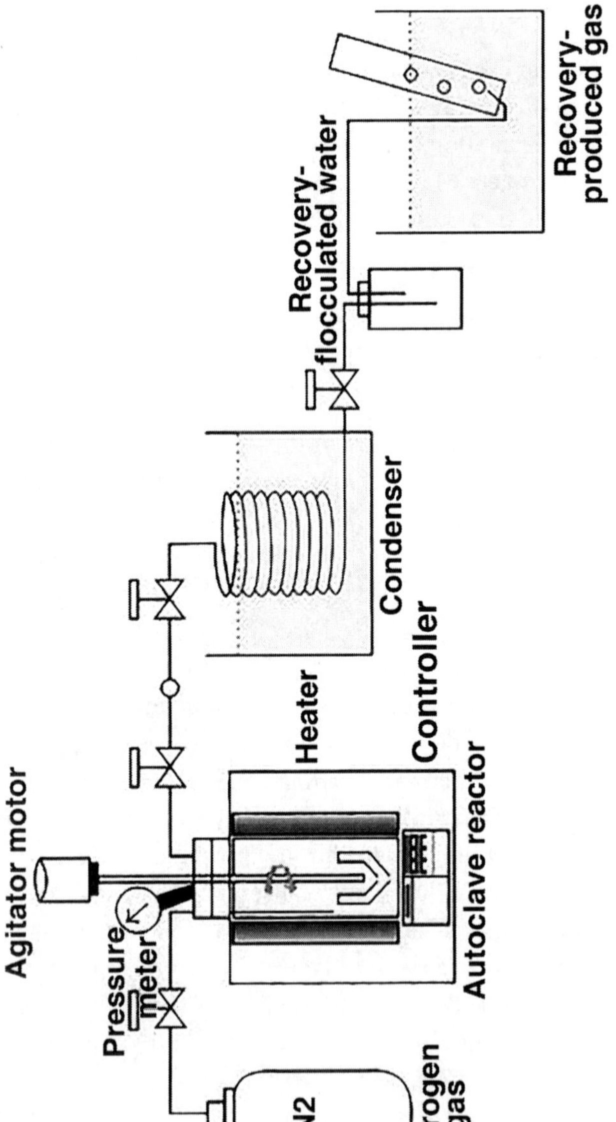

FIGURE 1: Schematic diagram of the autoclave facility.

13.3 EXPERIMENT METHODOLOGY

13.3.1 APPARATUS AND EXPERIMENTAL PROCEDURE

The hydrothermal treatment experiments were performed using the 500 mL autoclave facility as shown in Figure 1. The autoclave facility consists of a reactor, a heater, and a steam condenser which was operated with N_2 gas. For all experiments, the raw samples were milled so that their sizes become less than 1 cm for obtaining homogenous materials. The weight of one sample is 20 g which was mixed in the same amount of water and loaded into the reactor. The operating temperatures (pressures) of the hydrothermal treatment were 180°C (1.8 MPa), 200°C (2.0 MPa), and 220°C (2.4 MPa), and the reaction period was 30 minutes with the agitation speed of 200 rpm. It is reported that the subcritical hydrolysis starts at proximately 180°C of the reaction temperature [13]. After finishing the hydrothermal reaction, the pressure and the temperature fell down to atmospheric and room temperature, and the products were taken out of the reactor.

13.3.2 MATERIALS

The MT residue composition is based on the one obtained from Mokpo city, middle sized city in Republic of Korea. It is consisted of food residue (40–50%), paper (30–40%), plastic, wood, rubber, and others in negligible amount. In detail, the food residue is composed of various compositions such as vegetables (72%), fruits (15%), cereals, meat, and fish.

In this experiment, the food residue and paper contents were chosen as two parameters used, which are the highest ratio in the composition of the MT residue. Japanese newspaper and Korean kimchi in various compositions were used in substitute of paper and food residues, which were manually prepared by blending after the crushing process. In order to investigate the effect of main components in paper and kimchi, cellulose sample (α-Cellulose-fibriform from Nacalai Tesuque Inc., Kyoto, Japan) and lignin (Kanto Chemical Co., Inc., Japan) were also used for the hydrothermal treatment experiments.

13.3.3 ANALYSIS

The dehydration performance of raw sample and hydrothermal products was determined using a centrifugal separator with variable speed from 2,000 to 14,000 rpm. The natural drying tests to evaluate the moisture content reduction of the raw materials and the hydrothermally treated products after the centrifuge dehydration were conducted in the room temperature.

The ultimate analysis of the raw samples and solid products were carried out using the PerkinElmer made 2400 Series II CHN organic elemental analyzer. The proximate analyses were conducted using the SHIMADZU D-50 simultaneous TGA/DTA analyzer. The calorific values were measured using the bomb calorimetric method according to the JIS M-8814. SEM microphotographs were taken by JSM-6610LA analytical scanning electronic microscope after drying the solid products. The biomass composition measurement of raw paper and kimchi was entrusted to Nihon Hakko Shiryo Company. The biomass composition of cellulose, hemicelluloses, and lignin was defined in (1) to (3).

$$\text{Hemicellulose (\%)} = \text{NDF} - \text{ADF} \tag{1}$$

$$\text{Cellulose (\%)} = \text{ADF} - \text{ADL} \tag{2}$$

$$\text{Lignin (\%)} = \text{ADL} \tag{3}$$

where, NDF is neutral detergent fiber, ADF is acid detergent fiber, and ADL is acid detergent lignin.

13.4 RESULTS AND DISCUSSION

13.4.1 BIOMASS COMPOSITION OF RAW MATERIAL

Table 1 shows the biomass composition of the paper and kimchi utilized in the experiments. The biomass compositions of both raw materials were completely different. The paper was composed of 57.2% cellulose, 12.3% lignin, and 6.9% hemicelluloses whose total was 76.5%. The kimchi was composed of 16.8% cellulose, 4.6% lignin, and 0.5% hemicelluloses whose total is 21.9%. The biomass composition of raw material should influence the property of the hydrothermal treatment product. Therefore, in this research, cellulose and lignin samples were also used to compare with the paper and the kimchi.

TABLE 1: Biomass composition of paper and kimchi.

Biomass composition (wt. %), (d.b)	Paper	Kimchi	Calorific value (MJ/kg), (d.b)
Cellulose	57.26	16.78	16.5
Hemicellulose	6.95	0.51	16.7 [23]
Lignin	12.28	4.64	20.4
Others and Ash	23.51	78.07	—

d.b: dry basis.

Table 2 shows the proximate and ultimate analysis of the paper and kimchi utilized in the experiments. The kimchi had high moisture content of 92.4%, while the moisture content of paper was as low as 2.3%. High-efficiency moisture removal from the MT residue is one of purposes of this research. Volatile matter content of the paper and kimchi were as high as 87.0% and 67.1%, respectively, while the fixed carbon contents are low. The carbon contents in the paper and kimchi were 40.3% and 33.6%, respectively. The oxygen contents in the paper and kimchi were as high as 53.8% and 57.6%, respectively.

13.4.2 IMPROVEMENT OF DEHYDRATION AND DRYING PERFORMANCES OF KIMCHI BY THE HYDROTHERMAL TREATMENT

Removal of moisture contents in MSW is a major target of the pretreatment, and the moisture content of MSW has a strong influence on the characteristics and treatment method of MSW [5, 20–22].

TABLE 2: Proximate and ultimate analysis results of paper and kimchi and their products.

	Paper				Kimchi			
	Raw material	Treated at 180°C	Treated at 200°C	Treated at 220°C	Raw material	Treated at 180°C	Treated at 200°C	Treated at 220°C
Moisture (a.r)	2.3	4.0	4.2	6.5	92.4	93.8	93.2	93.3
Proximate analysis (d.b)								
Volatile matter	87.0	76.2	58.4	56.6	67.1	60.3	60.9	57.8
Fixed carbon	5.3	14.1	27.0	29.2	22.6	29.8	29.7	31.0
Ash	7.7	9.7	14.6	14.2	10.3	10.0	9.4	11.3
Ultimate analysis (wt. %) (d.a.f.)								
C	40.3	45.0	54.5	54.8	33.6	34.4	35.8	37.0
H	5.6	5.4	5.0	4.8	5.3	4.6	4.6	4.5
N	0.2	0.1	0.4	0.2	3.5	3.2	3.2	3.0
O	53.8	49.5	40.1	40.2	57.6	57.8	56.4	55.4
Composition of biomass (wt. %), (d.b)								
Cellulose	57.3	41.0	41.3	36.2	16.8	11.5	11.1	9.5
Hemicellulose	7.0	0.8	0.8	0.5	0.5	0.1	0.1	0.0
Lignin	12.3	13.6	14.1	14.7	4.6	4.9	4.9	5.0
Other and ash	23.5	44.6	43.9	48.6	78.1	83.5	84.0	85.5
Weight loss (wt. %), (d.b)		9.5	16.3	20.6		5.6	8.4	11.2
Calorific value (MJ/kg) (d.b)	15.5	16.9	21.4	21.7	14.7	14.7	16.0	15.8

a.r: as-received, d.b: dry basis, d.a.f: dry ash free.

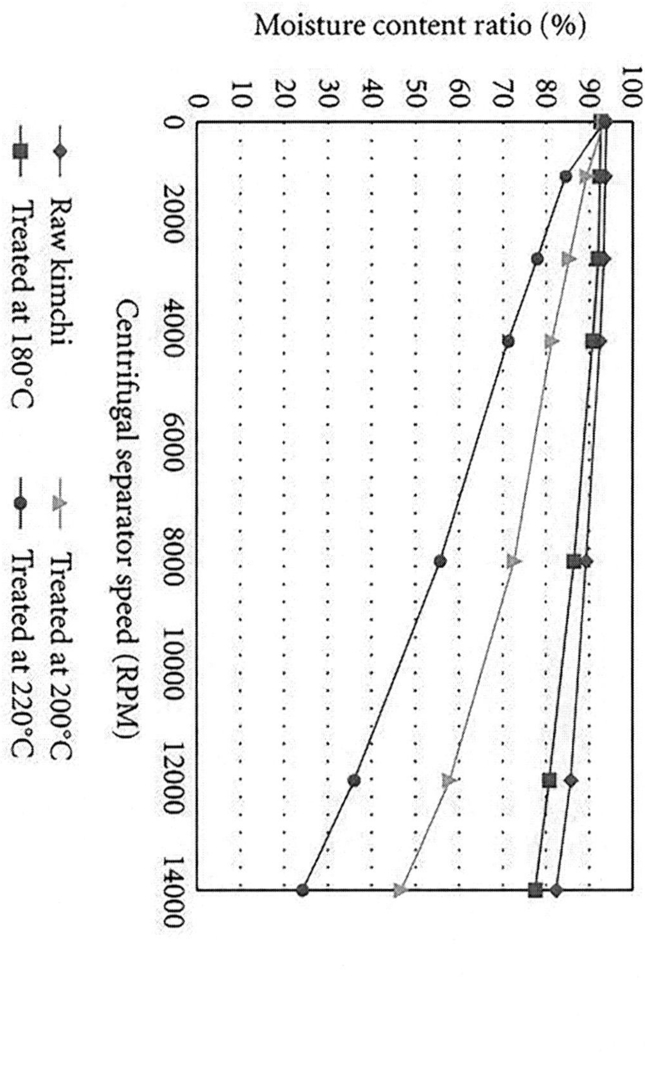

FIGURE 2: Dehydration performance of raw kimchi and its hydrothermal products.

Figure 2 shows the dehydration performance of raw kimchi and its hydrothermal products treated at 180°C, 200°C, and 220°C and then dehydrated by using a centrifugal separator with various rotation speeds from 2,000 to 14,000 rpm. As shown in Table 1 and Figure 2, raw kimchi had 92.5% moisture content. The moisture contents of the hydrothermal products in the experiments at 180°C, 200°C and 220°C were 93.8%, 93.2%, and 93.4%, respectively.

The hydrothermal products had better dehydration performance compared with raw kimchi, and the moisture content could be reduced down to 24.2% when treated at 220°C, with the centrifugal separator speed of 14,000 rpm. In the case of the hydrothermal treatment at 180°C, the color of the product was changed but the result of the dehydration performance was similar to raw kimchi.

Figure 3 shows the natural drying performance of the raw kimchi and the dehydrated residue of its hydrothermal products with the centrifugal speed of 14,000 rpm under the room temperature. It shows the time change of the moisture content of the dehydrated residue with different reaction temperatures. The moisture content of the residue treated at 200°C and 220°C decreased down to approximately 10% after 36 hours and 24 hours, respectively, while the moisture content of the raw kimchi and the dehydrated residue with the reaction temperature of 180°C could not be reduced below 30% by this natural drying within 70 hours.

The results clearly show that the hydrothermal treatment can improve the dehydration and drying performances of kimchi, which should lead in the reduction of energy requirement for the moisture removal from kimchi.

13.4.3 EFFECT OF THE HYDROTHERMAL TREATMENT ON THE PROPERTY OF THE PRODUCTS

Hydrothermal treatment breaks the physical and chemical structure in the materials such as cellulose, hemicelluloses, and lignin [10–14] in paper and kimchi, and these biomasses were broken down into smaller and simpler molecules.

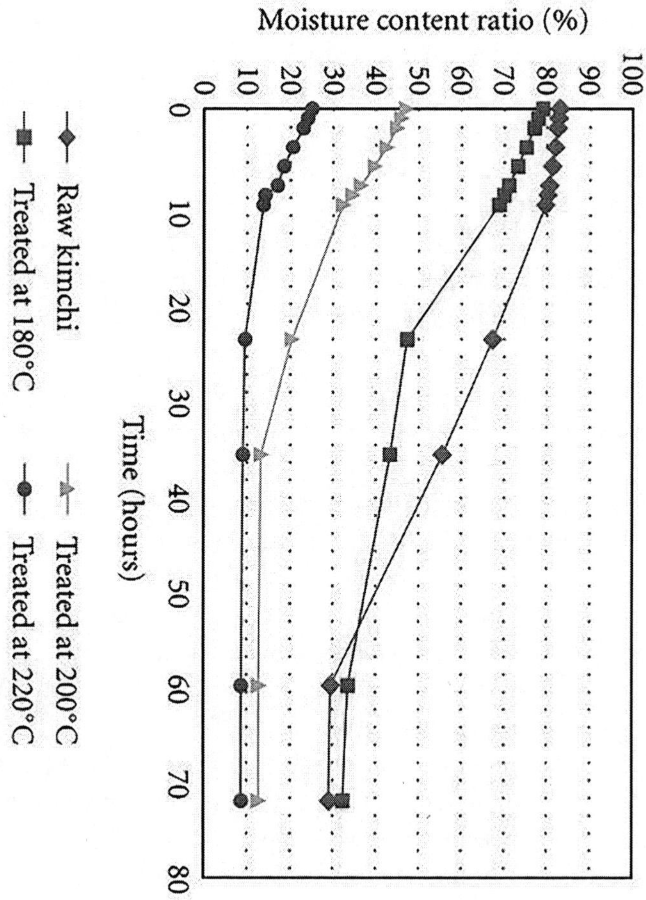

FIGURE 3: Natural drying performance of raw kimchi and, the dehydrated residue of its hydrothermal products.

Table 2 shows the property of raw samples of kimchi and paper and their products after the hydrothermal treatment which were produced at 180°C, 200°C, and 220°C. The chemical properties of the paper and the kimchi were changed by the hydrothermal treatment.

The paper and kimchi had high volatile matter content (87.0% and 67.1%) and oxygen content (53.8% and 57.6%) like other biomass. With the increase of the hydrothermal reaction temperature, the volatile matter and oxygen content decreased while the fixed carbon content increased, which were caused by the hydrolysis reaction.

Proximate and ultimate analysis of the paper and its hydrothermal products showed more significant change than the kimchi. The volatile matter of the paper decreased from 87.0% to 76.2%, 59.7%, and 56.9% at the reaction temperature of 180°C, 200°C, and 220°C, respectively.

Figure 4 shows the calorific value of the paper and the kimchi before and after the hydrothermal treatment, together with those of cellulose and lignin. The calorific values of the kimchi and paper increased with the increase of the reaction temperature due to the increase of the fixed carbon content, where this calorific value increase is more significant in the case of the paper than in the case of the kimchi.

As shown in Table 1, the paper was composed of 57.3% cellulose, 12.3% lignin, 7.0% hemicelluloses, and 23.5% others and ash, while the kimchi is composed of 16.8% of cellulose, 4.6% of lignin, 0.5% of hemicelluloses, and 78.1% of other and ash. Table 2 shows biomass compositions of the hydrothermal products which were treated at 180°C, 200°C and 220°C. About 40% and 60% cellulose decomposed at 200°C and 220°C, respectively. About 90% and 99% hemicellulose decomposed at 180°C and 220°C, respectively. Cellulose and hemicellulose were decomposed to be smaller molecular by the hydrolysis reaction in the hydrothermal treatment. However, the lignin behavior was different from cellulose and hemicellulose. When hydrothermal treated at 180 to 220°C, less than 5% of lignin was decomposed. The results suggest that cellulose and hemicellulose were easier to decompose than lignin. As discussed elsewhere [12, 14], lignin starts to decompose at the temperature exceeding 250°C. In addition, the hydrothermal treatment increases the calorific value of cellulose as shown in Figure 4 which explains the reason of the calorific value increase of the paper after the hydrothermal treatment whose major component is cellulose.

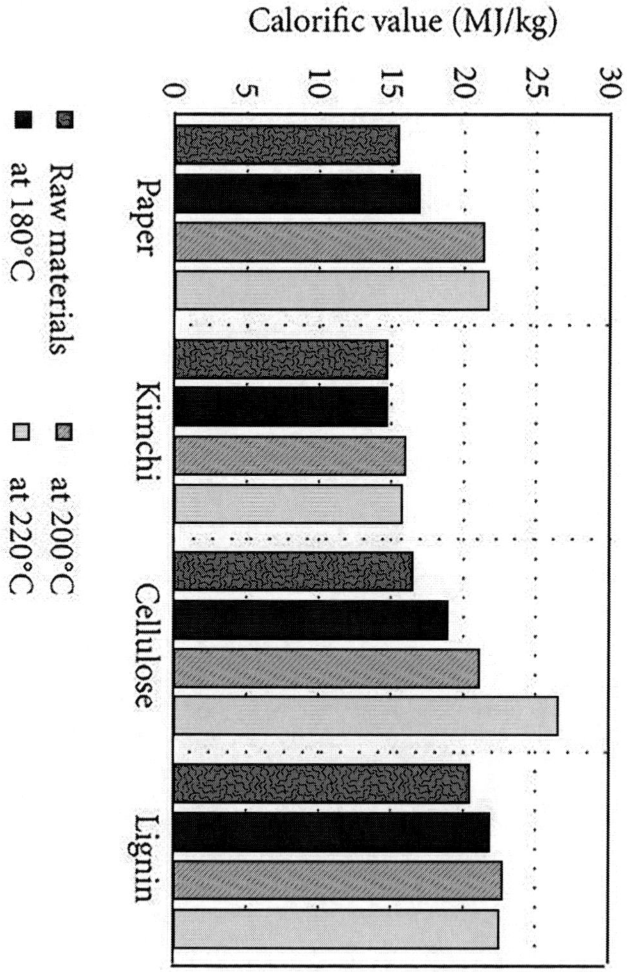

FIGURE 4: Effect of the hydrothermal treatment on the calorific value of products.

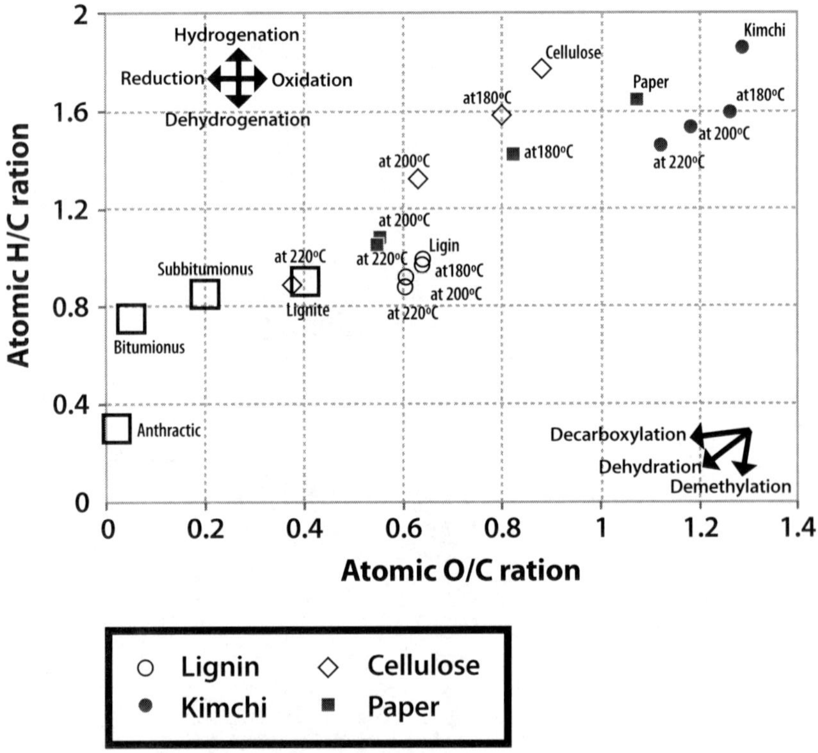

FIGURE 5: Coalification band of the paper and the kimchi at different hydrothermal reaction temperatures in comparison with cellulose, lignin and coal.

FIGURE 6: SEM microphotographs of the raw materials and their hydrothermal products; (a) paper and (b) kimchi.

The hydrothermal treatment changes the properties of products like the coalification process. Figure 5 shows the coalification band of the raw materials (paper, kimchi, cellulose, and lignin) and their hydrothermal products in comparison with various kinds of coal. The hydrothermal treatment can promote water removal from waste and biomass by improving the dewatering performance as well as the chemical dehydration. The chemical dehydration significantly increases the heating value by decreasing the H/C and O/C ratios. The chemical dehydration of cellulose can be expressed as $4(C_6H_{10}O_5)_n \leftrightarrow 2(C_{12}H_1O_5)_n + 10H_2O$ where the hydrothermal reaction promotes cleavage of mainly eater and chemical bonds of cellulose.

Paper and kimchi are known to have high atomic H/C ratios and atomic O/C ratios similar to other biomass [10, 12]. The atomic H/C ratio and the atomic O/C ratio were decreased by increasing the hydrothermal reaction temperature, where the paper and the cellulose showed more significant change of the atomic H/C ratio and the atomic O/C ratio compared with the kimchi and the lignin and should approach to lignite characteristic by increasing the hydrothermal reaction temperature. The fuel upgrading behavior of lignin is less significant in the hydrothermal reaction temperature range from 180°C to 220°C.

Figure 6 shows the SEM microphotographs of the paper and kimchi before and after the hydrothermal treatment. These SEM microphotographs reveal the changes between the raw materials and the upgraded solid products, showing disruption of physical structures and formation of individual grains in the products. Apparently, the hydrothermal treatment breaks the structure of the paper and kimchi and converts them into smaller uniform particle products.

13.4.4 EFFECT OF MIXING OF THE PAPER AND KIMCHI

If we can predict the effect of the hydrothermal treatment for the mixture of materials based on the hydrothermal behavior of individual material in the MT residue, it becomes much easier to evaluate the effectiveness of the hydrothermal treatment. Figures 7 and 8 show the change of the calorific value of the hydrothermal products by changing the mixing ratio of the paper and kimchi with various hydrothermal reaction temperatures with fixed hydrothermal reaction time of 30 minutes.

The blending ratios of paper to kimchi were from 100%:0% to 0%:100%. In these figures, the measured calorific values of the hydrothermal products of each mixture and the predicted calorific value of the hydrothermal products by the linear interpolation of the calorific value of the hydrothermal products of the paper and kimchi showed a good agreement. From these figures, we can see that the calorific value of the hydrothermal products of the mixture of the paper and kimchi can be well predicted based on the individual hydrothermal behavior, and the fuel upgrading by the hydrothermal treatment becomes more significant by increasing the amount of the paper.

13.5 CONCLUSION

In this research, the hydrothermal treatment was conducted for the paper, the kimchi and their mixture surrogating the Korean MT residue to demonstrate the improvement of dehydration and drying performances as well as fuel upgrading with the reaction temperature of 180°C, 200°C and 220°C, with the reaction time of 30 minutes. The dehydration and natural drying performances of the kimchi were significantly improved by the hydrothermal treatment. SEM microphotography showed that the physical structure of fibers of the paper and kimchi were broken into smaller and simpler molecules by the hydrothermal treatment.

In the case of paper, the volatile matter decreased from 87.0% to 58.4% and the fixed carbon increased from 5.3% to 27.0% at the reaction temperature of 200°C. As a result, the calorific value also increased from 14.7 MJ/kg to 21.7 MJ/kg at the reaction temperature of 200°C. On the other hand, this fuel upgrading behavior of the kimchi was rather weak due to its low cellulose content. As for the mixture of the paper and kimchi, the fuel upgrading behavior by the hydrothermal treatment was well predicted by the individual fuel upgrading behavior of the paper and kimchi.

These results demonstrated the effectiveness of the hydrothermal treatment of the MT residue for fuel upgrading.

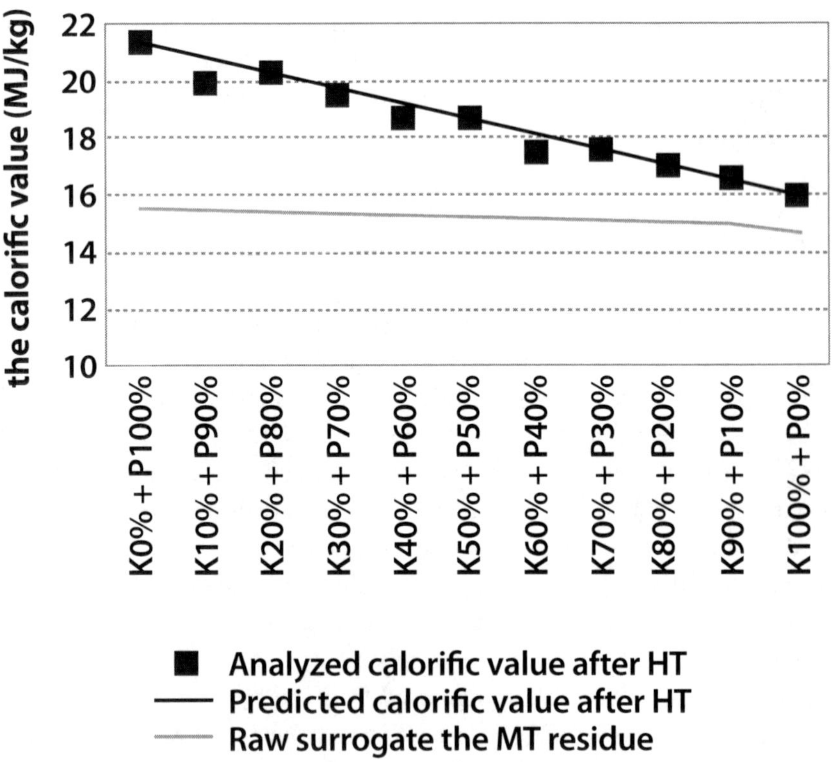

FIGURE 7: Change of the calorific value of the hydrothermal products by changing the mixture ratio of the paper andkimchi (hydrothermal reaction temperature = 200°C).

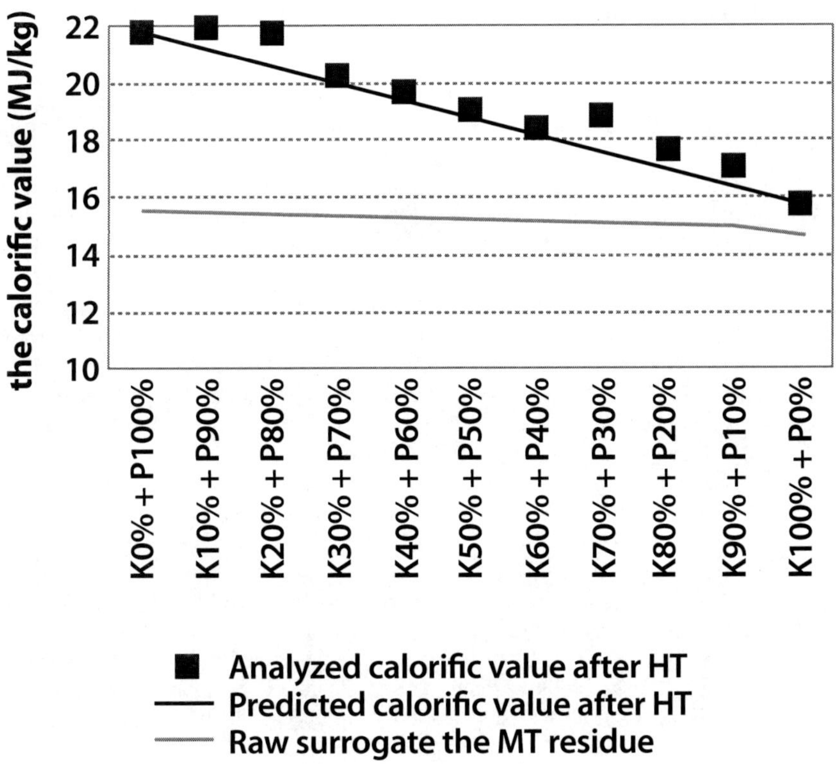

FIGURE 8: Change of the calorific value of the hydrothermal products by changing the mixture ratio of the paper and kimchi (hydrothermal reaction temperature = 220°C).

REFERENCES

1. Y. Kuzuhara, "Biomass Nippon strategy—why "biomass Nippon" now?" Biomass and Bioenergy, vol. 29, no. 5, pp. 331–335, 2005.
2. Juniper Consultancy Services Ltd, "Mechanical-Biological Treatment: A Guide for Decision makers/processes, policies and markets," 2005.
3. A. P. Economopoulos, "Technoeconomic aspects of alternative municipal solid wastes treatment methods," Waste Management, vol. 30, no. 4, pp. 707–715, 2010.
4. R. Bayard, J. de Araújo Morais, G. Ducom, F. Achour, M. Rouez, and R. Gourdon, "Assessment of the effectiveness of an industrial unit of mechanical-biological treatment of municipal solid waste," Journal of Hazardous Materials, vol. 175, no. 1-3, pp. 23–32, 2010.
5. Y. S. Yun, J. I. Park, and J. M. Park, "High-rate slurry-phase decomposition of food wastes: indirect performance estimation from dissolved oxygen," Process Biochemistry, vol. 40, no. 3-4, pp. 1301–1306, 2005.
6. R. Barrena, G. d'Imporzano, S. Ponsá et al., "In search of a reliable technique for the determination of the biological stability of the organic matter in the mechanical-biological treated waste," Journal of Hazardous Materials, vol. 162, no. 2-3, pp. 1065–1072, 2009.
7. K. Suksankraisorn, S. Patumsawad, and B. Fungtammasan, "Co-firing of Thai lignite and municipal solid waste (MSW) in a fluidised bed: effect of MSW moisture content," Applied Thermal Engineering, vol. 30, no. 17-18, pp. 2693–2697, 2010.
8. G. Luo, X. Cheng, W. Shi, P. James Strong, H. Wang, and W. Ni, "Response surface analysis of the water:feed ratio influences on hydrothermal recovery from biomass," Waste Management, vol. 31, pp. 438–444, 2011.
9. J. C. Serrano-Ruiz, D. J. Braden, R. M. West, and J. A. Dumesic, "Conversion of cellulose to hydrocarbon fuels by progressive removal of oxygen," Applied Catalysis B, vol. 100, no. 1-2, pp. 184–189, 2010.
10. A. Funke and F. Ziegler, "Hydrothermal carbonization of biomass: a summary and discussion of chemical mechanisms for process engineering," Biofuels, Bioproducts and Biorefining, vol. 4, no. 2, pp. 160–177, 2010.
11. P. Krammer and H. Vogel, "Hydrolysis of esters in subcritical and supercritical water," Journal of Supercritical Fluids, vol. 16, no. 3, pp. 189–206, 2000.
12. O. Bobleter, "Hydrothermal degradation of polymers derived from plants," Progress in Polymer Science, vol. 19, no. 5, pp. 797–841, 1994.
13. M. Sakaguchi, K. Laursen, H. Nakagawa, and K. Miura, "Hydrothermal upgrading of Loy Yang Brown coal—effect of upgrading conditions on the characteristics of the products," Fuel Processing Technology, vol. 89, no. 4, pp. 391–396, 2008.
14. A. T. Yuliansyah, T. Hirajima, S. Kumagai, and K. Sasaki, "Production of solid biofuel from agricultural wastes of the palm oil industry by hydrothermal treatment," Waste and Biomass Valorization, vol. 1, pp. 395–405, 2010.
15. S. M. Heilmann, H. T. Davis, L. R. Jader et al., "Hydrothermal carbonization of microalgae," Biomass and Bioenergy, vol. 34, no. 6, pp. 875–882, 2010.

16. P. Prawisudha, T. Namioka, and K. Yoshikawa, "Coal alternative fuel production from municipal solid wastes employing hydrothermal treatment," Applied Energy, vol. 90, no. 1, pp. 298–304, 2012.
17. A. T. Mursito, T. Hirajima, K. Sasaki, and S. Kumagai, "The effect of hydrothermal dewatering of Pontianak tropical peat on organics in wastewater and gaseous products," Fuel, vol. 89, no. 12, pp. 3934–3942, 2010.
18. A. T. Mursito, T. Hirajima, and K. Sasaki, "Upgrading and dewatering of raw tropical peat by hydrothermal treatment," Fuel, vol. 89, no. 3, pp. 635–641, 2010.
19. S. M. Heilmann, H. T. Davis, L. R. Jader et al., "Hydrothermal carbonization of microalgae," Biomass and Bioenergy, vol. 34, no. 6, pp. 875–882, 2010.
20. A. Hammerschmidt, N. Boukis, E. Hauer et al., "Catalytic conversion of waste biomass by hydrothermal treatment," Fuel, vol. 90, no. 2, pp. 555–562, 2011.
21. M. Nonaka, T. Hirajima, and K. Sasaki, "Upgrading of low rank coal and woody biomass mixture by hydrothermal treatment," Fuel, vol. 90, no. 8, pp. 2578–2584, 2011.
22. S. Luo, B. Xiao, Z. Hu, S. Liu, and X. Guo, "An experimental study on a novel shredder for municipal solid waste (MSW)," International Journal of Hydrogen Energy, vol. 34, no. 3, pp. 1270–1274, 2009.
23. E. Panayotova-Björnbom, P. Björnbom, J.-C. Cavalier, and E. Chornet, "The combined dewatering and liquid phase hydrogenolysis of raw peat using carbon monoxide," Fuel Processing Technology, vol. 2, no. 3, pp. 161–169, 1979.

Author Notes

CHAPTER 1

Acknowledgments

The Authors wish to thank José Diez and Kristyna Stiovova from Maguin S.A.S. and Lat Grand Ndiaye for their valuable help in the collection of the samples. This research project was included in the "Institut de Recherche en ENvironnement Industriel" (IRENI) and was financially supported by the Nord-Pas-de-Calais Regional Council, the French Ministry of Higher Education and Research, the European Regional Development Funds (through the Regional Delegation for Research and Technology).

CHAPTER 2

Acknowledgments

This study was supported by Zhejiang province Natural Science Foundation (Y5100192).

CHAPTER 3

Acknowledgments

The authors gratefully acknowledge Veolia (F. GELIX, D. BORRUT, F. NICOL, B. LEFEBVRE) for supplying the test samples and for partial financial support to complete this project. In addition, the team help of the Combustion and Catalysis Laboratory is gratefully acknowledged, especially Dr. H. C. BUTTERMAN who completed the experimental work.

CHAPTER 5

The paper has been prepared within the frame of the National Science Centre project based on decision no DEC-2011/03/D/ST8/04035.

Conflicts of Interest

The authors declare no conflict of interest.

CHAPTER 7

Acknowledgments

This research work was financially supported by National Key Technology R & D Program in the 11th Five year Plan of China (Grant No. 2008BAC46B06).

CHAPTER 9

Conflict of Interest

The authors declare that there is no conflict of interests regarding the publication of this paper.

Acknowledgments

The authors would like to thank the Ministry of New and Renewable Energy (MNRE, India) and the University Grant Commission (UGC) for financial support and the management committee and the staff of the N. Shankaran Nair Research Center (NSNRC, Ambernath, India) for carrying out experiments and providing laboratory facilities.

CHAPTER 10

Acknowledgments

The authors thank the support of Program for New Century Excellent Talents in University (NCET-10-0529).

CHAPTER 11

Conflict of Interest

The authors declare that there is no conflict of interests regarding the publication of this paper.

Acknowledgments

The authors gratefully acknowledge Council of Scientific & Industrial Research (CSIR), India, for their financial support.

CHAPTER 12

Acknowledgments

Su Shiung Lam acknowledges the financial assistance by Public-Service-Department of Malaysia Government and University Malaysia Terengganu.

Index

D

N

O

P